Report of Investigations 9676

Explosion Effects on Mine Ventilation Stoppings

By Eric S. Weiss, Kenneth L. Cashdollar, Samuel P. Harteis, P.E.,
Gary J. Shemon, Dennis A. Beiter, and John E. Urosek

DEPARTMENT OF HEALTH AND HUMAN SERVICES
Centers for Disease Control and Prevention
National Institute for Occupational Safety and Health
Pittsburgh Research Laboratory
Pittsburgh, PA

November 2008

This document is in the public domain and may be freely copied or reprinted.

Disclaimer

Mention of any company or product does not constitute endorsement by the National Institute for Occupational Safety and Health (NIOSH). In addition, citations to Web sites external to NIOSH do not constitute NIOSH endorsement of the sponsoring organizations or their programs or products. Furthermore, NIOSH is not responsible for the content of these Web sites. All Web addresses referenced in this document were accessible as of the publication date.

Ordering Information

To receive documents or other information about occupational safety and health topics, contact NIOSH at

>Telephone: **1–800–CDC–INFO** (1–800–232–4636)
>TTY: 1–888–232–6348
>e-mail: cdcinfo@cdc.gov
>
>or visit the NIOSH Web site at **www.cdc.gov/niosh**.

For a monthly update on news at NIOSH, subscribe to NIOSH *eNews* by visiting **www.cdc.gov/niosh/eNews**.

DHHS (NIOSH) Publication No. 2009–102

November 2008

SAFER • HEALTHIER • PEOPLE™

CONTENTS

Page

Abstract 1
Introduction 2
Experimental mine and test procedures 3
Stopping construction 6
    Concrete block stoppings 6
    Steel panel stoppings 14
    Australian woven cloth stoppings 21
Explosion test results 26
    Dry-stacked hollow-core concrete block stoppings in crosscuts 26
        Test #427 27
        Test #428 28
        Test #429 38
        Test #430 38
        Test #432 38
        Test #433 42
        Test #434 46
    Steel panel stoppings in crosscuts 50
        Test #457 50
        Test #458 50
    Dry-stacked solid-concrete-block stoppings in crosscuts 54
        Test #457 54
        Test #458 54
        Test #459 56
        Test #460 57
        Test #461 58
        Test #462 58
        Test #463 62
    Wet-laid solid-concrete-block stoppings in crosscuts 65
        Test #510 65
        Tests #511–#514 65
        Test #515 65
        Tests #516–#518 67
        Test #519 67
    Dry-stacked hollow-core concrete block stoppings in entry and crosscuts 70
        Test #491 72
        Test #494 72
        Test #495 72
        Test #496 72
        Test #497 74
    Australian woven cloth stoppings in crosscuts 80
        Test #459 80
        Test #460 80
        Test #461 85
        Test #462 85

CONTENTS—Continued

Page

Discussion and conclusions 89
Acknowledgments 91
References 92
Appendix.—Summary tables of explosion pressure data for the LLEM tests 94

ILLUSTRATIONS

1. Plan view of the Lake Lynn experimental mine 4
2. Plan view of the LLEM showing the multiple-entry area and stopping locations during LLEM tests #457 and #458 5
3. Dry-stacked hollow-core concrete block stopping construction, with staggered joints 7
4. Dry-stacked hollow-core concrete blocks used in construction of the stoppings 8
5. Installation of wood wedges between the stopping and rib for dry-stacked hollow-core concrete block stopping 8
6. Installation of wood header boards and wood wedges between the stopping and roof for dry-stacked hollow-core concrete block stopping 9
7. Coating the dry-stacked hollow-core concrete block stopping with sealant 9
8. Dry-stacked solid-concrete-block stopping construction with staggered joints 10
9. Wood wedges between the stopping and rib for dry-stacked solid-concrete-block stopping ... 11
10. Wood header boards and wood wedges between the stopping and roof 11
11. Completed dry-stacked solid-concrete-block stopping in X-5 as viewed from C-drift, with the total explosion pressure transducer at the upper right 12
12. Wet-laid solid-concrete-block stopping construction with staggered joints 13
13. Completed 6-in (15-cm) thick, wet-laid solid-concrete-block stopping in X-7, as viewed from A-drift 13
14. Horizontal steel angles for attaching the steel stopping panels 15
15. Three horizontal steel angles inset into rib 15
16. Horizontal steel angles (with open sides facing panels on left), as viewed from B-drift 16
17. Polystyrene foam head seal used to provide enhanced closure between metal panel and roof, as viewed from the C-drift side 17
18. Temporary jack used to exert roof-to-floor pressure on the metal panel during installation ... 17
19. Metal twist clamp used to secure open side of panel to the horizontal angle 18
20. Overlap panel used to close gap between rib panel and second panel, as viewed from the front or C-drift side 18
21. Side extensions used to fill gaps between rib panel and rib, viewed from the back or B-drift side 19
22. Polyurethane sealant used on perimeter and vertical panel joints 20
23. Nearly completed steel panel stopping in X-6 showing closed, smooth panel sides facing C-drift 20
24. Nearly completed steel panel stopping in X-6 showing the open panel side attached to the horizontal rib-to-rib angles on the B-drift side 21
25. Inserting pipe into top sleeve of woven cloth 23
26. Inserting pipe/sleeve into top box section 23

CONTENTS—Continued

Page

27. Anchoring top box section to the mine roof 24
28. Wedging woven cloth into mine rib 24
29. Sealant along perimeter of the stopping, viewed from C-drift 25
30. Rib bolts installed in the X-6 stopping, viewed from B-drift 25
31. Completed woven cloth stopping design in X-6, viewed from the explosion or C-drift side 26
32. Plan view of the multiple-entry area of LLEM showing effects of explosion test #428 29
33. LVDT displacement, pressure-time integrals, and pressures versus time at dry-stacked hollow-core concrete block stopping in X-4 during LLEM test #428 30
34. Debris from dry-stacked hollow-core concrete block stopping in X-4 after LLEM test #428, viewed from C-drift 31
35. Expanded map of debris from the X-4 dry-stacked hollow-core concrete block stopping after LLEM test #428 31
36. LVDT displacement, pressure-time integrals, and pressures versus time at the dry-stacked hollow-core concrete block stopping in X-5 during LLEM test #428 32
37. Debris from X-5 dry-stacked hollow-core concrete block stopping after LLEM test #428, viewed from C-drift 33
38. Expanded map of debris from the X-5 dry-stacked hollow-core concrete block stopping after LLEM test #428 33
39. Remnants of the dry-stacked hollow-core concrete block stopping in X-5 after LLEM test #428 34
40. LVDT displacement, pressure-time integrals, and pressures versus time at the dry-stacked hollow-core concrete block stopping in X-6 during LLEM test #428 35
41. Dry-stacked hollow-core concrete block stopping in X-6 remained intact after LLEM test #428 36
42. LVDT displacement, pressure-time integrals, and pressures versus time at the dry-stacked hollow-core concrete block stopping in X-7 during LLEM test #428 36
43. Pressures at walls (ribs) and at dry-stacked hollow-core concrete block stopping locations during LLEM test #428 37
44. Expanded cross-sectional views of the blocks remaining in X-6 after test #430. 39
45. Crosscut 6 dry-stacked hollow-core concrete block stopping viewed from B-drift side after LLEM test #430, showing blocks dislodged by explosion 40
46. Closeup of B-drift side of X-6 dry-stacked hollow-core concrete block stopping after test #430, showing dislodged and damaged blocks 40
47. Closeup of C-drift side X-6 dry-stacked hollow-core concrete block stopping after test #430, showing missing and damaged blocks 41
48. The dry-stacked hollow-core concrete block stopping in X-7 remained intact after LLEM test #430 41
49. Expanded cross-sectional view of the blocks remaining in X-6 after test #433 42
50. Debris from X-6 dry-stacked hollow-core concrete block stopping after LLEM test #433, viewed from C-drift 43
51. Part of the inby side of X-6 dry-stacked hollow-core concrete block stopping remaining after LLEM test #433, viewed from C-drift 43

CONTENTS—Continued

Page

52. Expanded map of debris from the X-6 dry-stacked hollow-core concrete block stopping after LLEM test #433 44
53. Expanded cross-sectional view of the blocks remaining in X-7 after explosion test #433, viewed from C-drift side 44
54. Condition of X-7 dry-stacked hollow-core concrete block stopping after explosion test #433, viewed from C-drift side 45
55. Expanded view of top part of X-7 dry-stacked hollow-core concrete block stopping after explosion test #433, showing damaged blocks 45
56. Expanded view of hollow-core concrete block that is almost dislodged from X-7 dry-stacked stopping after explosion test #433 46
57. Effects of explosion test #434 on the dry-stacked hollow-core concrete block stoppings, with expanded cross-sectional views of the blocks remaining in X-6 and X-7 47
58. Additional blocks partially dislodged from the right side of X-6 dry-stacked hollow-core concrete block stopping after test #434, viewed from C-drift 48
59. Crosscut 7 dry-stacked hollow-core concrete block stopping viewed from B-drift after LLEM test #434 48
60. Crosscut 7 dry-stacked hollow-core concrete block stopping viewed from C-drift after LLEM test #434 49
61. Map of debris from the X-7 dry-stacked hollow-core concrete block stopping after LLEM test #434 49
62. LVDT displacement, pressure-time integrals, and pressures versus time at the steel panel stopping in X-6 during LLEM test #458 51
63. Debris from steel panel stopping in X-6 after LLEM test #458, viewed from C-drift 52
64. Map of debris from the X-6 steel panel stopping after LLEM test #458 52
65. LVDT displacement, pressure-time integrals, and pressures versus time at the steel panel stopping in X-7 during LLEM test #458 53
66. Debris from steel panel stopping in X-7 after LLEM test #458, viewed from B-drift 53
67. Closeup of steel panel stopping still partially attached to outby rib in X-7 after LLEM test #458, viewed from B-drift 54
68. LVDT displacement, pressure-time integrals, and pressures versus time at the dry-stacked solid-concrete-block stopping in X-4 during LLEM test #458 55
69. LVDT displacement, pressure-time integrals, and pressures versus time at the dry-stacked solid-concrete-block stopping in X-4 during LLEM test #459 56
70. Displacement of the second block course on the X-4 dry-stacked solid-concrete-block stopping after LLEM test #459, viewed from C-drift 57
71. LVDT displacement, pressure-time integrals, and pressures versus time at the dry-stacked solid-concrete-block stopping in X-4 during LLEM test #462 59
72. Debris from the dry-stacked solid-concrete-block stopping and remaining blocks on the outby rib line in X-4 after LLEM test #462, viewed from C-drift 59
73. Debris from the dry-stacked solid-concrete-block stopping and remaining blocks on the inby rib line in X-4 after LLEM test #462, viewed from C-drift 60
74. Map of debris from the X-4 dry-stacked solid-concrete-block stopping after LLEM test #462 60

CONTENTS—Continued

Page

75. LVDT displacement, pressure-time integrals, and pressures versus time at the dry-stacked solid-concrete-block stopping in X-5 during LLEM test #462 61
76. Pressures at walls (ribs) and at the dry-stacked solid-concrete-block stopping locations during LLEM test #462 62
77. LVDT displacement, pressure-time integrals, and pressures versus time at the dry-stacked solid-concrete-block stopping in X-5 during LLEM test #463 63
78. Remaining blocks from the dry-stacked solid-concrete-block stopping at the inby rib in X-5 after LLEM test #463, viewed from B-drift 64
79. Debris map for the X-5 dry-stacked solid-concrete-block stopping after LLEM test #463 64
80. LVDT displacement, pressure-time integrals, and pressures versus time at the 8-in (20-cm) thick, wet-laid solid block stopping in X-6 during LLEM test #515 66
81. LVDT displacement, pressure-time integrals, and pressures versus time at the 6-in (15-cm) thick, wet-laid solid block stopping in X-7 during LLEM test #515 67
82. LVDT displacement, pressure-time integrals, and pressures versus time at the 8-in (20-cm) thick, wet-laid solid block stopping in X-6 during LLEM test #519 68
83. LVDT displacement, pressure-time integrals, and pressures versus time at the 6-in (15-cm) thick, wet-laid solid block stopping in X-7 during LLEM test #519 69
84. Debris from 6-in (15-cm) thick, wet-laid solid block stopping in X-7 after LLEM test #519, viewed from A-drift 69
85. Completed dry-stacked hollow-core concrete block stopping with simulated open regulator across B-drift 71
86. Dry-stacked hollow-core concrete block stopping between B- and C-drifts in X-4 showing individually numbered block, viewed from C-drift 71
87. Debris map after LLEM test #491 73
88. Pressure-time integrals and pressures versus time at the dry-stacked hollow-core concrete block stopping across B-drift during LLEM test #497 74
89. Debris from the dry-stacked hollow-core concrete block stopping across B-drift after LLEM test #497 75
90. Pressure-time integrals and pressures versus time at the X-4 dry-stacked hollow-core concrete block stoppings during LLEM test #497 75
91. Damage to the X-4 dry-stacked hollow-core concrete block stopping between B- and C-drifts after LLEM test #497 76
92. Debris from the X-4 dry-stacked hollow-core concrete block stopping between A- and B-drifts after LLEM test #497 76
93. Pressure-time integrals and pressures versus time at the X-3 dry-stacked hollow-core concrete block stoppings during LLEM test #497 77
94. Damage to the X-3 dry-stacked hollow-core concrete block stopping between B- and C-drifts after LLEM test #497 77
95. Debris map after LLEM test #497 78
96. Total explosion pressure versus time at the dry-stacked hollow-core concrete block stoppings in X-4 and in B-drift during LLEM test #497 79
97. Pressure-time integrals and pressures at the X-6 woven cloth stopping during LLEM test #459 81

CONTENTS—Continued

Page

98. Pressure-time integrals and pressures at the X-7 woven cloth stopping during LLEM test #459 82
99. Pressure-time integrals and pressures at the X-6 woven cloth stopping during LLEM test #460 82
100. Pressure-time integrals and pressures at the X-7 woven cloth stopping during LLEM test #460 83
101. Tears in the woven cloth on the outby rib of X-6 stopping after LLEM test #460, viewed from C-drift 84
102. Initial tear in the woven cloth at the corner steel roof bolt plate position on the inby rib of the X-7 stopping after LLEM test #460, viewed from C-drift 84
103. Pressure-time integrals and pressures at the X-6 woven cloth stopping during LLEM test #462 86
104. Pressure-time integrals and pressures at the X-7 woven cloth stopping during LLEM test #462 87
105. Damage to the woven cloth stopping in X-6 after LLEM test #462 88
106. Damage to the woven cloth stopping in X-7 after LLEM test #462 88
107. Results of the LLEM block stopping evaluations compared to the critical design parameters and the transverse load capacity predictions of the NIOSH empirical correlation by Barczak and Batchler (for dry-stacked block walls) and WAC (for wet-laid block walls) 90

TABLES

1. Evaluations of mine stoppings during the LLEM explosion tests 27
2. Peak total explosion pressure data at dry-stacked hollow-core concrete block stoppings for test #428 in C-drift 38
3. Peak total explosion pressure data at dry-stacked solid-concrete-block stoppings for test #462 in C-drift 62
4. Peak total explosion pressure data at wet-laid solid-concrete-block stoppings for test #510 in A-drift 65
5. Peak total explosion pressure data at wet-laid solid-concrete-block stoppings for test #515 in A-drift 66
6. Peak total explosion pressure data at wet-laid solid-concrete-block stoppings for test #519 in A-drift 70
7. Peak total explosion pressure data at Australian woven cloth stoppings for test #460 in C-drift 83
8. Peak total explosion pressure data at Australian woven cloth stoppings for test #462 in C-drift 85
9. Total explosion pressures necessary to destroy typical U.S. stoppings in the LLEM 89
A-1. Maximum explosion pressures during LLEM test #427 94
A-2. Maximum explosion pressures during LLEM test #428 94
A-3. Maximum explosion pressures during LLEM test #429 95
A-4. Maximum explosion pressures during LLEM test #430 95

CONTENTS—Continued

Page

A-5. Maximum explosion pressures during LLEM test #432 96
A-6. Maximum explosion pressures during LLEM test #433 96
A-7. Maximum explosion pressures during LLEM test #434 97
A-8. Maximum explosion pressures during LLEM test #457 97
A-9. Maximum explosion pressures during LLEM test #458 98
A-10. Maximum explosion pressures during LLEM test #459 98
A-11. Maximum explosion pressures during LLEM test #460 99
A-12. Maximum explosion pressures during LLEM test #461 99
A-13. Maximum explosion pressures during LLEM test #462 100
A-14. Maximum explosion pressures during LLEM test #463 100
A-15. Maximum explosion pressures during LLEM test #491 101
A-16. Maximum explosion pressures during LLEM test #494 101
A-17. Maximum explosion pressures during LLEM test #495 102
A-18. Maximum explosion pressures during LLEM test #496 102
A-19. Maximum explosion pressures during LLEM test #497 103
A-20. Maximum explosion pressures during LLEM test #510 103
A-21. Maximum explosion pressures during LLEM test #512 104
A-22. Maximum explosion pressures during LLEM test #515 104
A-23. Maximum explosion pressures during LLEM test #519 105

ACRONYMS AND ABBREVIATIONS USED IN THIS REPORT

CFR	Code of Federal Regulations
DG	data-gathering
LLEM	Lake Lynn Experimental Mine
LVDT	linear variable displacement transducer
MSHA	Mine Safety and Health Administration
NIOSH	National Institute for Occupational Safety and Health
PRL	Pittsburgh Research Laboratory (NIOSH)
WAC	Wall Analysis Code
X	crosscut (e.g., "X-1" stands for "crosscut 1")

UNIT OF MEASURE ABBREVIATIONS USED IN THIS REPORT

cm	centimeter
ft	foot
ft^2	square foot
ft^3	cubic foot
g/m^3	gram per cubic meter
in	inch
km	kilometer
kPa	kilopascal
kPa-s	kilopascal second
m	meter
m^2	square meter
m^3	cubic meter
mm	millimeter
MPa	megapascal
ms	millisecond
psi	pound-force per square inch
psi-s	pound-force per square inch - second
sec	second

EXPLOSION EFFECTS ON MINE VENTILATION STOPPINGS

By Eric S. Weiss,[1] Kenneth L. Cashdollar,[2] Samuel P. Harteis,[3] Gary J. Shemon,[4] Dennis A. Beiter,[5] and John E. Urosek[6]

ABSTRACT

The National Institute for Occupational Safety and Health (NIOSH) and the Mine Safety and Health Administration (MSHA) conducted joint research to evaluate explosion blast effects on typical U.S. mine ventilation stoppings in the NIOSH Pittsburgh Research Laboratory's (PRL) Lake Lynn Experimental Mine (LLEM). An innovative Australian-designed brattice stopping was also evaluated.

After mine explosion accidents, MSHA conducts investigations to determine the cause(s) as a means to prevent future occurrences. As part of these postexplosion investigations, the condition of underground stoppings, including the debris from damaged stoppings, is documented as evidence of the approximate strength and the direction of the explosion forces. The LLEM data showed that a typical dry-stacked and coated solid-concrete-block stopping survived a total explosion pressure of ~6.7 psi (~46 kPa) and was destroyed at a total explosion pressure of ~7.6 psi (~52 kPa). In comparison, a typical dry-stacked and coated hollow-core concrete block stopping survived a total explosion pressure of ~3.4–4.3 psi (~23–30 kPa) and was destroyed at a total explosion pressure of ~3.6–5.2 psi (~25–36 kPa), depending on the length of the pressure pulse and the value of the pressure-time integral. A typical steel panel stopping design survived a total explosion pressure of 0.8 psi (5.5 kPa) and failed at a total explosion pressure of 1.3 psi (9 kPa). The LLEM data also showed that an obstacle blocking the path of a pressure wave resulted in a higher reflected pressure at the obstacle. An 8-in (20-cm) thick wet-laid solid-concrete-block stopping coated on one side survived a total explosion pressure of ~26 psi (~180 kPa); this stopping was not tested to failure. A 6-in (15-cm) thick wet-laid solid-concrete-block stopping coated on one side survived a total explosion pressure of ~14 psi (~97 kPa) and was destroyed at a total explosion pressure of ~25 psi (~172 kPa). An innovative Australian woven cloth stopping survived an explosion pressure of 4.0 psi (27 kPa) and was destroyed at an explosion pressure of ~6.1 psi (~42 kPa). These results will help investigators determine the approximate explosion forces that destroy or damage stoppings during actual coal mine explosions.

---

[1]Team Leader (Senior Research Mining Engineer), Lake Lynn Laboratory Section, Pittsburgh Research Laboratory, National Institute for Occupational Safety and Health, Pittsburgh, PA.

[2]Principal Research Physical Scientist, Pittsburgh Research Laboratory, National Institute for Occupational Safety and Health, Pittsburgh, PA.

[3]Research Mining Engineer, Pittsburgh Research Laboratory, National Institute for Occupational Safety and Health, Pittsburgh, PA.

[4]Mining Engineer, Mine Ventilation and Emergency Services Branch, Ventilation Division, Pittsburgh Safety and Health Technology Center, Mine Safety and Health Administration, Triadelphia, WV.

[5]Supervisory Mining Engineer, Mine Ventilation and Emergency Services Branch, Ventilation Division, Pittsburgh Safety and Health Technology Center, Mine Safety and Health Administration, Triadelphia, WV.

[6]Chief, Mine Emergency Operations, Pittsburgh Safety and Health Technology Center, Mine Safety and Health Administration, Pittsburgh, PA.

INTRODUCTION

Permanent stoppings are used to control and direct the ventilation airflow through underground coal mines to dilute and render harmless methane, entrained coal dust, and other contaminants at the working face and other areas of the mine. 30 CFR[7] 75.333 requires that permanent stoppings be built and maintained between intake and return air courses beginning at the third connecting crosscut outby the working face and to separate other air courses and direct air as specified. To perform the intended function and meet the requirements of 30 CFR 75.333, permanent stoppings are to be constructed in a traditionally accepted method and of materials that have been demonstrated to perform adequately or in a method and of materials that have been tested and shown to have a minimum strength equal to or greater than the traditionally accepted in-mine controls. A few examples of traditionally accepted [61 Fed. Reg.[8] 9764 (1996)] stopping construction methods are as follows: (1) 8-in (20-cm) and 6-in (15-cm) concrete block (both hollow-core and solid) with mortared joints, (2) 8-in (20-cm) and 6-in (15-cm) concrete blocks, dry-stacked and coated on one or both sides with a strength-enhancing sealant suitable for dry-stacked stoppings, and (3) steel stoppings (minimum 20-gauge) with seams and perimeter sealed with a suitable mine sealant.

Unlike mine ventilation seal structures [30 CFR 75.335; Greninger et al. 1991; Mitchell 1971; Weiss et al. 1993, 1996, 1999, 2002] that are commonly used to isolate unused sections of the mine, stoppings are not intended to withstand explosion overpressures. Unfortunately, mine explosions do occur. Depending on the location and severity, explosions can result in fatalities and injuries to underground mining personnel and cause considerable underground damage to equipment and structures. In the mine explosions in Alabama in 2001 and West Virginia in 2006, ventilation stoppings were destroyed [McKinney et al. 2002; Gates et al. 2007]. Mine Safety and Health Administration (MSHA) personnel conduct investigations into these types of explosion accidents to determine the root cause(s) as a means to prevent future occurrences. As part of postexplosion investigations, the location and condition of underground ventilation structures and debris are mapped. This information helps the investigators determine the strength and the direction of the forces of the explosion.

Previous research by Kawenski et al. [1965] conducted in the Bruceton Experimental Mine on the strength of stoppings revealed that a 16-ft (4.9-m) wide by 6-ft (1.8-m) high dry-stacked, concrete block stopping constructed with hollow-core block, wedged at the roof, and coated on the low-pressure side was ruptured when subjected to a 2.5-psi (17-kPa) pressure pulse generated from the burning of black powder. A similar concrete block stopping with hollow-core block but with fully mortared joints, no wedges, and coated on the low-pressure side ruptured at about 5.8 psi (40 kPa). The hollow-core concrete blocks used during these evaluations at Bruceton were cinder block (8 in by 8 in by 16 in (20 cm by 20 cm by 40 cm)) for some stopping designs and solid gravel block (8 in by 6 in by 16 in (20 cm by 15 cm by 40 cm)) for others. As expected, this research also showed that the pressure required for rupture decreased as the stopping width increased.

The more recent stopping evaluations in the Lake Lynn Experimental Mine (LLEM) involved various full-scale designs subjected to known overpressures generated from methane and/or coal dust explosions. These stopping designs were typical of those currently used in U.S. underground coal mines. In addition to evaluating the effects of the explosion on typical

---

[7] *Code of Federal Regulations.* See CFR in references.
[8] *Federal Register.* See Fed. Reg. is references.

dry-stacked and coated concrete block stoppings located in crosscuts (perpendicular to the flow of the pressure pulse), a similar study was conducted with a stopping installed across the entry outby from where the explosion was initiated. This study was designed to simulate actual explosion accidents in coal mines where some crosscut stoppings near the ignition location survived the initial explosion pressure pulse, but stoppings farther away were destroyed. Wet-laid solid-concrete-block stoppings coated on one side were also evaluated to determine the enhanced explosion overpressure resistance provided by the mortared joints. In addition, in cooperation with the Australian mining industry, a new and innovative woven cloth stopping proposed for use in Australian mines was evaluated. The construction and testing methods, explosion test data, and postexplosion condition of each stopping design are presented in this report.

In addition, the results from the numerous stopping evaluations conducted in the LLEM were compared to predictions using a NIOSH empirical correlation for dry-stacked concrete block walls [Barczak 2005; Barczak and Batchler 2006, 2008] and the Wall Analysis Code (WAC) for wet-laid concrete block walls developed by the U.S. Army Corps of Engineers [Slawson 1995].

EXPERIMENTAL MINE AND TEST PROCEDURES

The explosion evaluation tests on the coal mine stopping designs were conducted at the NIOSH Lake Lynn Laboratory [Mattes et al. 1983; Triebsch and Sapko 1990]. Lake Lynn Laboratory is a multipurpose research laboratory designed to provide a modern, full-scale, realistic environment for conducting surface and underground research in mining health and safety technology. Lake Lynn Laboratory is located about 50 miles (80 km) southeast of Pittsburgh, near Fairchance, Fayette County, PA, and occupies a former limestone mine. It is one of the world's foremost mining laboratories for conducting large-scale surface and underground health and safety research.

The underground LLEM (Figure 1) is unique in that it can simulate current U.S. coal mine geometries for a variety of mining scenarios, including multiple-entry room-and-pillar mining and longwall mining. The old limestone mine workings are shown on the left in Figure 1. Five new drifts (horizontal passageways in a mine) were developed to simulate the geometries of modern U.S. coal mines. LLEM has four parallel drifts: A, B, C, and D. D-drift is a 1,640-ft (500-m) long single entry that can be separated from E-drift by an explosion-proof bulkhead door. To simulate room-and-pillar workings, drifts A, B, and C can be used. These three drifts are each approximately 1,600 ft (490 m) long, with seven crosscuts at the inby end. An explosion-proof bulkhead door is used to separate the multiple entries from E-drift. Drifts C and D are connected by E-drift, a 500-ft (152-m) long entry that simulates a longwall face. Explosion tests can be conducted in the single-entry D-drift; the multiple-entry area of A-, B-, and C-drifts; or various other configurations including the longwall E-drift. The entries are about 20 ft (6 m) wide by about 6.5 ft (2 m) high, with cross-sectional areas of 130–140 ft^2 (12–13 m^2). From August 1983 (when the first explosion test was conducted) to July 2008, a total of 527 consecutively numbered explosion tests were conducted in the LLEM.

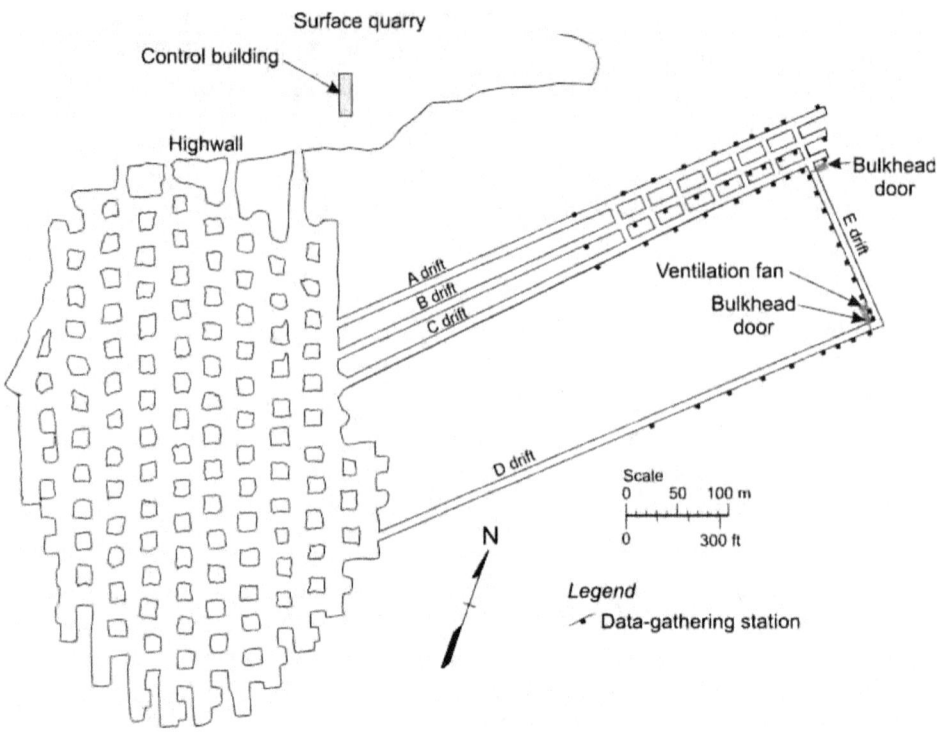

Figure 1.—Plan view of the Lake Lynn Experimental Mine.

Figure 2 shows an expanded view of the stopping test area in the multiple-entry section of the LLEM. The faces, or closed ends, of A-, B-, and C-drifts are on the right in the figure. Most of the explosion evaluations were conducted in C-drift and the stoppings were built in the crosscuts between B- and C-drifts, as shown in Figure 2. Note that, in the LLEM, the first crosscut ("1" in Figure 2) is the one nearest the face. During four of the evaluations, the permanent stoppings were constructed in crosscut 4 (X-4)[9] at 355 ft (108 m), X-5 at 451 ft (138 m), X-6 at 547 ft (167 m), or X-7 at 647 ft (197 m) from the face of C-drift. The crosscuts are 17–19 ft (5.2–5.8 m) wide by ~7.2 ft (~2.2 m) high with a cross-sectional area of about 130 ft^2 (12 m^2). Explosion-resistant seals from a previous study were located in X-1 through X-3. The head-on explosion pressure evaluation in B-drift included stoppings in X-3 and X-4 between A- and B-drifts and between B- and C-drifts and one stopping across B-drift between X-4 and X-5. The evaluation of the wet-laid stoppings was conducted in A-drift as part of another explosion program with the stoppings located in X-6 and X-7 between A- and B-drifts; there were seals located in X-1 through X-5 between A- and B-drifts. These last two scenarios are not shown in Figure 2. Before each explosion test, a 60-ton pneumatically operated, track-mounted, concrete and steel bulkhead was positioned across E-drift to contain the explosion pressures within the multiple-entry area. The LLEM bulkhead door and some of the other infrastructure were designed to withstand explosion overpressures of up to 100 psi (690 kPa). Higher pressures have been recorded at areas away from these structures.

---

[9]The abbreviation "X" stands for "crosscut" throughout this report, e.g., "X-1" stands for "crosscut 1."

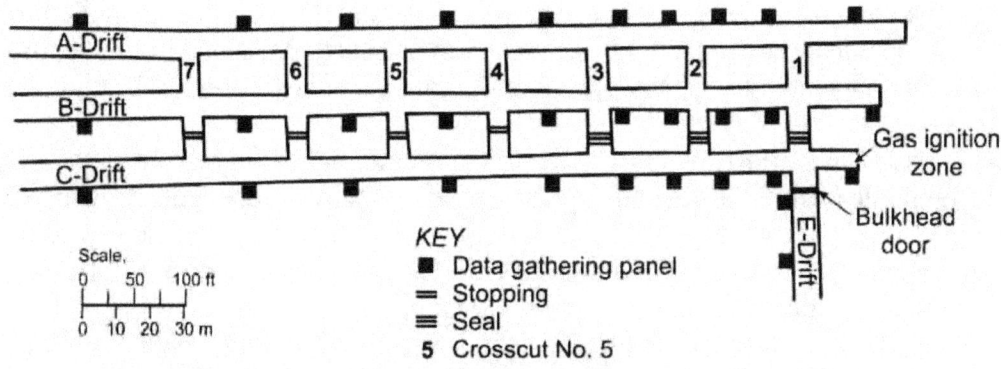

Figure 2.—Plan view of the LLEM showing the multiple-entry area and stopping locations during LLEM tests #457 and #458.

For the LLEM explosion tests, natural gas was injected into the ignition zone. This natural gas is composed of ~97%–98% methane, ~1.5% ethane, and small percentages of other higher hydrocarbons. Sample lines within the ignition zone were used to draw gas samples to an infrared analyzer on the surface for continuous monitoring of the methane concentration. In addition, samples were collected in evacuated test tubes and sent to verify the analyses using gas chromatography. Most of the tests for these stopping failure evaluations used a ~9% methane-air concentration within a 10-ft (3-m) deep by 11-ft (3.4-m) wide ignition zone (~760 ft^3 (~21.5 m^3)) contained in the C-drift face area with a clear plastic diaphragm. A few of the tests used a larger gas ignition zone. A fan with an explosion-proof motor housing mixed the natural gas and air prior to ignition. Electrically activated matches located either at the face (closed end) or outby the face within the gas ignition zone, depending on the explosion overpressure desired, were used to ignite the methane-air mixtures. In some of the tests, shelves of pulverized bituminous coal dust were located at a distance of 10–40 ft (3–12 m) from the face as a means to increase the explosion overpressures. For each of these explosion tests, the gas was ignited and the explosion pressure traveled out C-drift. In addition, five of the explosion tests were conducted in B-drift to evaluate the effect of an obstruction in the path of an oncoming pressure wave. During these B-drift tests, the ~6%–10% methane-air concentration was contained within a 20-ft (6-m) or 40-ft (12-m) long section of entry located between X-2 and X-3. For the explosion tests conducted in A-drift, the length of the gas ignition zone was varied to obtain higher total explosion overpressures at the stopping locations, i.e., the ~10% methane-air concentration was contained within a 40-, 50-, or 85-ft (12-, 15-, or 26-m) long gas ignition zone (as measured from the closed end of A-drift) for the different tests.

Each LLEM drift has 10 data-gathering (DG) stations inset in the rib wall at the locations shown in Figures 1 and 2. Each DG station houses a strain gauge transducer to measure the explosion pressure and an optical sensor to detect the flame arrival. The explosion pressure is dynamic in nature and is composed of two components: an omnidirectional pressure component (the pressure that is exerted in all directions and is measured perpendicular to the gas flow; also referred to by others as a "quasi-static pressure") and a wind or velocity pressure component (pressure due to gas flow). The total explosion pressure is the sum of the omnidirectional pressure and the wind or velocity pressure. The transducers in the DG stations in the wall measure the omnidirectional pressure. All of the explosion pressures presented in this report are overpressures or gauge pressures (pressures above local atmospheric pressure) rather than absolute pressures.

In the past, Nagy [1981, p. 58] referred to the omnidirectional pressure as the "static pressure" to differentiate it from the "dynamic pressure," or velocity component. However, the omnidirectional pressure is not actually "static" as it does vary with time during the explosion. This terminology by Nagy had been used in previous U.S. Bureau of Mines and NIOSH publications, but it was confusing to many readers and will no longer be used.

The pressure transducers of particular interest for these stopping evaluations were along the C-drift rib at 304, 403, 501, 598, and 757 ft (92.7, 122.8, 152.7, 182.3, and 230.7 m); along the B-drift rib at 257, 329, and 427 ft (78.3, 100.3, and 130.2 m); and along the A-drift rib at 550, 649, and 807 ft (167.6, 197.8, and 246.0 m) as measured from the respective C-, B-, and A-drift faces. There were also pressure transducers mounted on the C-drift side of each of the stoppings when the explosion was initiated in C-drift, on the B-drift side of the stoppings when the explosion was initiated in B-drift, and on the A-drift side of the stoppings when the explosion was initiated in A-drift. These transducers faced the explosion forces traveling into the crosscuts and measured the total explosion pressures at the stoppings. The velocity (or wind) pressure would act on objects in an open entry as the explosion travels down the entry. When the velocity (or wind) pressure reaches a solid surface, such as a stopping across the entry or crosscut, the velocity would go to zero and the omnidirectional pressure at that location becomes approximately the same as the total explosion pressure. In addition, when the pressure pulse reaches an obstruction across the entry (such as a stopping in B-drift for one of these evaluations), it is reflected and the resulting total reflected pressure can be about twice the incoming pressure pulse value.

Attached at the center (midheight and midwidth) of the nonexplosion side of the concrete block and steel panel stoppings was a linear variable differential transducer (LVDT) that measured the movement of the center of the stopping during each explosion. A photograph and details of the operation of the LVDT are presented by Weiss et al. [1999, pp. 5–6]. For the C-drift tests, the spring-loaded LVDT was mounted around a 90° bend on the B-drift rib line outside of the crosscut and connected to the stopping via lightweight, near zero stretch fishing line. This mounting system protected the expensive LVDT from flying debris from the stopping. For the explosions initiated in B-drift, the LVDTs were not used. For the A-drift tests, the LVDTs were mounted within the crosscut behind a heavy steel angle designed to protect the sensor from flying debris; the sensor was connected to the stopping via the fishing line. During the explosion tests, a high-speed, PC-based National Instruments data acquisition system collected the data from the various instruments at a sampling rate of 1,500 per sec. The reported data were averaged over 10 ms using a 15-point smoothing technique to filter the higher frequency data.

STOPPING CONSTRUCTION

Concrete Block Stoppings

For the first series of tests, four 6-in (15-cm) thick dry-stacked hollow-core concrete block stoppings were constructed in X-4 through X-7 between C- and B-drifts in the LLEM (Figure 2). Crosscuts 4 and 5 were about 19 ft (5.8 m) wide by 7.2 ft (2.2 m) high, and crosscuts 6 and 7 were about 17 ft (5.2 m) wide by 7.2 ft (2.2 m) high at the positions of the stoppings.

Each stopping was located approximately 5 ft (1.5 m) toward B-drift from the midpoint of the crosscut (approximately 23 ft (7 m) deep into the crosscut, as measured from closest C-drift rib). The blocks were three-core concrete blocks, with nominal dimensions of 8 in by 6 in by 16 in (20 cm by 15 cm by 40 cm). The nominal uniaxial compressive strength of the block material was 1,900 psi (13 MPa). Full block testing by Barczak and Batchler [2008] revealed a hollow-core concrete block compressive strength of less than 1,000 psi (7 MPa). There was an 8-in (15-cm) thick reinforced concrete floor in the crosscuts where the stoppings were installed. To level the floor, a small concrete foundation that tapered from 0- to 3-in (0- to 8-cm) thick as measured from C- to B-drift was installed along the width of each crosscut, and a small amount of mortar was used under the first course of block to assist in the leveling of the first course of block. The remaining blocks were dry-stacked (no mortar between the block joints) with staggered joints (Figure 3). A closeup view of the dry-stacked hollow-core concrete blocks is shown in Figure 4. Wood wedges were used to tighten each block course at the mine ribs (Figure 5). Wood header boards and wedges were used between the top block course and the mine roof to tighten the structure from both sides of the stopping (Figure 6). An approximately 0.5-in (6-mm) thick coating of an approved sealant (Quikrete B-Bond, product No. 1227-50) was applied to both sides of the stoppings, as shown in Figure 7. The application of coatings on dry-stacked stoppings varies widely in practice. Some mines have coated their stoppings only on the high-ventilation pressure side, some have fully coated the high-ventilation pressure side with only a perimeter coating on the other side, and other mines have fully coated both sides.

Figure 3.—Dry-stacked hollow-core concrete block stopping construction, with staggered joints.

Figure 4.—Dry-stacked hollow-core concrete blocks used in construction of the stoppings.

Figure 5.—Installation of wood wedges between the stopping and rib for dry-stacked hollow-core concrete block stopping.

Figure 6.—Installation of wood header boards and wood wedges between the stopping and roof for dry-stacked hollow-core concrete block stopping.

Figure 7.—Coating the dry-stacked hollow-core concrete block stopping with sealant.

After completing the explosion test evaluation and removing the first set of dry-stacked hollow-core concrete block stoppings, two additional 6-in (15-cm) thick dry-stacked concrete block stoppings were constructed in X-4 and X-5 using solid concrete blocks, with nominal dimensions of 8 in by 6 in by 16 in (20 cm by 15 cm by 40 cm). Full block testing by Barczak and Batchler [2008] revealed a solid-concrete-block compressive strength of 1,330–1,780 psi (9–12 MPa). These stoppings were constructed in the same manner as the previously evaluated hollow-core concrete block stoppings. The solid concrete blocks were dry-stacked (no mortar between the block joints) with staggered joints (Figures 8–9). Wood wedges were used to tighten each block course at the mine ribs (Figures 9–10). Header boards were used between the top block course and the mine roof, and wedges were used between the header boards and mine roof to tighten the structure (Figures 9–10). An approximately 0.25-in (6-mm) thick coating of an approved sealant (Quikrete B-Bond, product No. 1227-50) was applied to both sides of the stoppings. The completed dry-stacked solid-concrete-block stopping in X-5, covered with sealant, is shown in Figure 11, with a pressure transducer suspended from the roof in the upper-right side of the photo.

Figure 8.—Dry-stacked solid-concrete-block stopping construction with staggered joints.

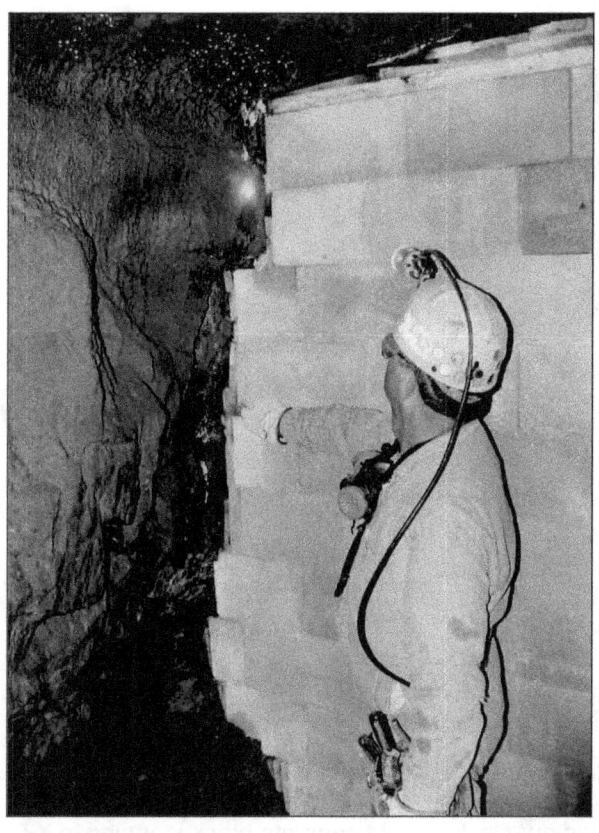

Figure 9.—Wood wedges between the stopping and rib for dry-stacked solid-concrete-block stopping.

Figure 10.—Wood header boards and wood wedges between the stopping and roof.

Figure 11.—Completed dry-stacked solid-concrete-block stopping in X-5 as viewed from C-drift, with the total explosion pressure transducer at the upper right.

Two additional solid-concrete-block stoppings were constructed in X-6 and X-7 between A- and B-drifts. These stoppings were constructed in the same manner as the previously evaluated solid-concrete-block stoppings in C-drift, except these stoppings were not dry-stacked. As can be seen in Figure 12, a full mortar bed (wet-laid) of Type S mortar was used in all of the horizontal and vertical block joints, including the perimeter gaps to the mine roof, ribs, and floor. Wood wedges were not used during the construction of these stoppings. An approximately 0.25-in (6-mm) thick coating of Quikrete's B-Bond was applied to the A-drift face (high-ventilation pressure side) of each stopping. The high-ventilation pressure side of a stopping may not necessarily be the side subjected to an explosion in an actual coal mine. All of the stoppings discussed in this report were coated on both sides or just on the high-ventilation pressure side, which was then also the side subsequently subjected to an explosion within the LLEM. Stoppings coated only on the nonexplosion side were not evaluated in the LLEM. The completed wet-laid solid-concrete-block stopping in X-7 is shown in Figure 13.

Figure 12.—Wet-laid solid-concrete-block stopping construction with staggered joints.

Figure 13.—Completed 6-in (15-cm) thick, wet-laid solid-concrete-block stopping in X-7, as viewed from A-drift.

Steel Panel Stoppings

Jack Kennedy Steel Stoppings, manufactured by Jack Kennedy Metal Products & Buildings, Inc., (hereinafter referred to as "steel panel stoppings"), were constructed within the LLEM per the manufacturer's instruction guide.[10] Both of the steel panel stoppings were installed, with assistance from NIOSH Lake Lynn staff and by personnel from Jack Kennedy Metal Products & Buildings, Inc., and its distributor, Ken-Air, Inc. These stoppings consisted of a series of 12-in (30-cm) wide by 72-in (183-cm) high by 2-in (5-cm) thick vertical telescoping steel panels (formed from 20-gauge galvanized steel sheeting) that could be lengthened or shortened to accommodate roof heights between 6 and 10 ft (1.8 and 3 m).

The stoppings were constructed on a small concrete foundation that tapered from 0- to 3-in (0- to 8-cm) thick along the width of X-6 and X-7 for leveling purposes. Each stopping was located approximately 5 ft (1.5 m) toward B-drift from the midpoint of the crosscut (or approximately 23 ft (7 m) deep into the crosscut, as measured from closest C-drift rib). Since the LLEM installation required the use of panel heights in excess of 5 ft (1.5 m) (MP-61020 kit, which can accommodate entry heights from 6 to 10 ft (1.8 to 3 m)), three rows of steel angle bars, extending horizontally from one rib to the other, were required, in accordance with the manufacturer's instruction guide (Figure 14). The angle bars were on the back or B-drift side of the steel panel stopping. The 1.5-in (3.8-cm) by 1.5-in (3.8-cm) by 12-ft (3.7-m) long hot-dipped galvanized steel angle bars (0.125 in (3 mm) thick) were positioned into small holes (approximately 2-in (5-cm) deep) that were cut into each rib (Figure 15). Two 12-ft (3.7-m) long lengths of angle were required per row to span the approximately 20-ft (6-m) wide crosscuts; the ~4-ft (~1.2-m) length of extra angle was merely overlapped into the other piece and taped in place. The angle bars were located at 22, 39, and 68 in (56, 99, and 173 cm) from the mine floor. As shown in Figure 16, the open end of the horizontal steel angles faced C-drift so as to rest against the channels of each vertical panel (on the back or open side of the panel). The front or smooth side of the panel (the closed side of the panel facing away from the horizontal rib-to-rib angles) was designed to face the entry with the highest ventilation pressure, i.e., toward C-drift for the LLEM evaluations.

---

[10]Contact the manufacturer for additional information on construction techniques and/or to obtain a detailed instruction guide.

Figure 14.—Horizontal steel angles for attaching the steel stopping panels. The photo also shows the installed center panel.

Figure 15.—Three horizontal steel angles inset into rib.

Figure 16.—Horizontal steel angles (with open sides facing panels on left), as viewed from B-drift.

To start the steel panel installation, the first panel was installed near the center of the crosscut (Figure 14). As part of the installation, a head seal (polystyrene foam) was manually inserted at the top and bottom of each panel to provide a better seal with the roof and floor (Figures 17–18). For coal mine applications, the head seal is generally only inserted at the top of the panel. However, for the LLEM application, a head seal was also used at the bottom of the panel to provide a better seal to the concrete mine floor. Then, a specially designed installation jack was positioned within the panel's top and bottom grooves and used to exert a roof-to-floor pressure until the jack started to slightly bow (per the manufacturer's instruction guide) to temporarily hold the stopping panel in place (similar to that shown in Figure 18). A wire twist clamp (Figure 19) was fastened around the horizontal rib-to-rib angle, inserted into the inside flanges of the panel, and then tightened. Six clamps were used on the panel to attach it to the three horizontal angles. The installation jack was then removed from the panel. The steel angles and clamps maintain compression and keep the panels aligned.

After installing the center panel, a second panel was installed against the outby rib, jacked into place, and secured with twist clamps. The panels were alternated such that the intersection of the telescoping panel sections offset each other. This was accomplished by turning every other panel upside down. The panel installation process was repeated until enough panels were added to reach the center panel, at which point the clamps on the center panel were loosened and that panel was slid over to the adjacent panel. Additional panels continued to be added to the point of being within 3–4 ft (~1 m) of the opposite rib. At that point, a panel was installed against the (opposite) rib and additional panels were then added back toward the other previously installed panels until the gap was ≤1 ft (≤30 cm). An overlap panel (indicated by an "X" in Figure 20) was used to cover this and other gaps between panels, and side extensions were used to cover any large gaps between the panels and the mine rib (Figure 21).

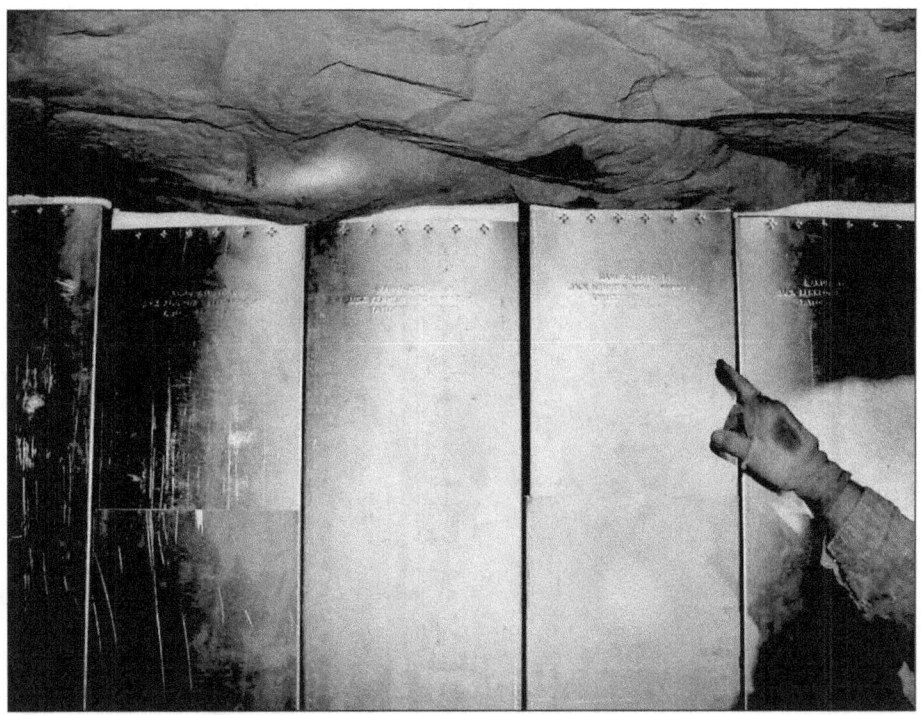

Figure 17.—Polystyrene foam head seal used to provide enhanced closure between metal panel and roof, as viewed from the C-drift side.

Figure 18.—Temporary jack used to exert roof-to-floor pressure on the metal panel during installation.

Figure 19.—Metal twist clamp used to secure open side of panel to the horizontal angle.

Figure 20.—Overlap panel (indicated by an "X") used to close gap between rib panel and second panel, as viewed from the front or C-drift side.

Figure 21.—Side extensions used to fill gaps between rib panel and rib, viewed from the back or B-drift side.

The stopping was sealed with a polyurethane VERSI-FOAM system (manufactured by RHH Foam Systems, Inc., New Berlin, WI, and distributed by Ken-Air, Inc., as the MP-567 sealant foam pack). The sealant was applied to the perimeter on both sides of the stopping and between all of the vertical panel joints on the smooth, or closed, panel side, which is the high-ventilation pressure side in an actual coal mine installation (Figure 22). All of the steel panel stoppings were constructed such that the high-ventilation pressure side of each stopping was on the C-drift side, which was the side that would be directly impacted by the explosion forces. Figure 23 shows the nearly completed steel panel stopping in X-6 as shown from the C-drift side (explosion or high-pressure side). Figure 24 shows the same stopping from the B-drift side.

Figure 22.—Polyurethane sealant used on perimeter and vertical panel joints.

Figure 23.—Nearly completed steel panel stopping in X-6 showing closed, smooth panel sides facing C-drift.

Figure 24.—Nearly completed steel panel stopping in X-6 showing the open panel side attached to the horizontal rib-to-rib angles on the B-drift side.

Australian Woven Cloth Stoppings[11]

Two Flexi-Stop stopping designs were constructed in X-6 and X-7 between C- and B-drifts in the LLEM after completing the explosion evaluations of the steel panel stoppings. (There were already seals in X-1 through X-3 and concrete block stoppings in X-4 and X-5.) The stopping in X-6 was located approximately 6 ft (1.8 m) toward C-drift as measured from the center of the crosscut or approximately 15 ft (4.6 m) into the crosscut from the closest C-drift rib. The X-7 stopping was approximately 8 ft (2.4 m) toward C-drift from the center or approximately 12 ft (3.7 m) into the crosscut. Crosscuts 6 and 7 were about 17 ft (5.2 m) wide and 7.2 ft (2.2 m) high. This report will only briefly summarize the basic construction materials and techniques; detailed information on the stopping construction can be obtained from representatives of Minova Australia.

The stopping designs consisted of a woven high-strength polymeric fiber coated with a fire-resistant product. Note that this Flexi-Stop stopping design as constructed and evaluated within the LLEM may not meet the fire-resistant qualities required for use in U.S. mines. The Flexi-Stop cloth was precut to the appropriate length and width based on the area where it was to be installed. A 0.8-in (2-cm) diameter galvanized steel pipe (rigid conduit with a ~0.1-in

---

[11]The woven cloth stopping evaluations were funded by Minova Australia.

(~0.25-cm) wall thickness) the length of the crosscut width was then installed through a heavily reinforced sewn sleeve along the top and bottom sections of the cloth (Figure 25). The woven cloth, with pipes installed, was then attached to each of the two box steel sections by means of a slot provided along the entire length of steel on one side of each box section. This resulted in the sleeve-pipe curtain assembly being inside the steel box section (Figure 26). The slot on the box section was large enough to allow the passage of the woven cloth but much smaller than the diameter of the pipe within the woven cloth sleeve, thereby providing a means to anchor the woven cloth to the roof and floor. The 2.5-in by 2.5-in (6.4-cm by 6.4-cm) steel box section used in X-6 had a wall thickness of 0.25 in (6 mm) and a slot width of approximately 0.5 in (1.3 cm). The 2-in by 2-in (5-cm by 5-cm) steel box section used in X-7 had a wall thickness of 0.1875 in (5 mm) and a slot width of about 0.3 in (8 mm). The assembly of these components was completed within the LLEM. Six steel ~3-in (~7.5-cm) long rings (~1.5-in (~3.8-cm) I.D. pipe for the X-6 frame and ~1.6-in (~4.1-cm) I.D. pipe for the X-7 frame), welded to one side of each steel box section, provided a means to anchor the box section to the mine roof and floor through the use of roof bolts. The steel rings that are attached to the box frame can be seen in Figure 27. The top and bottom box sections, with the woven cloth attached, were then bolted to the mine roof and floor using 1-in (2.5-cm) diameter by 26-in (66-cm) long resin bolts embedded 22 in (56 cm) into 1.375-in (3.5-cm) diameter drill holes (Figure 27). The woven cloth was intentionally oversized so as not to create a pretensioned surface. The woven cloth was then anchored to each rib by wedging the cloth (using wood wedges) into a 0.75-in (1.9-cm) wide by 8-in (20-cm) deep slot in each rib that extended from roof to floor (Figure 28). A small quantity of shotcrete was applied by trowel to a few of the larger gaps between the top box section in each crosscut and the roof. Minova's Tekflex sealant was then applied to the entire stopping perimeter from the C-drift side of each stopping (Figure 29). Following the first explosion test, 1-in (2.5-cm) diameter by 16-in (40-cm) long resin bolts were installed through the woven cloth from the B-drift side and into each rib (embedded 12 in (30 cm)) on the C-drift side of the rib slots. Standard steel 6-in by 6-in by 0.25-in (15-cm by 15-cm by 0.64-cm) thick roof bolt plates were used in conjunction with the bolts. An oversized piece of conveyor belt (R-11) was used under the bolt plate to protect the woven cloth from tearing. Five equally spaced bolts were used on each rib for the X-6 stopping and four bolts on each rib for the X-7 stopping. Figure 30 shows the installed rib bolts as seen from the B-drift side (nonexplosion side) of the X-6 stopping. Figure 31 shows the completed X-6 stopping as seen from the C-drift or explosion side, with the pressure transducer suspended from the roof in the center of the photo.

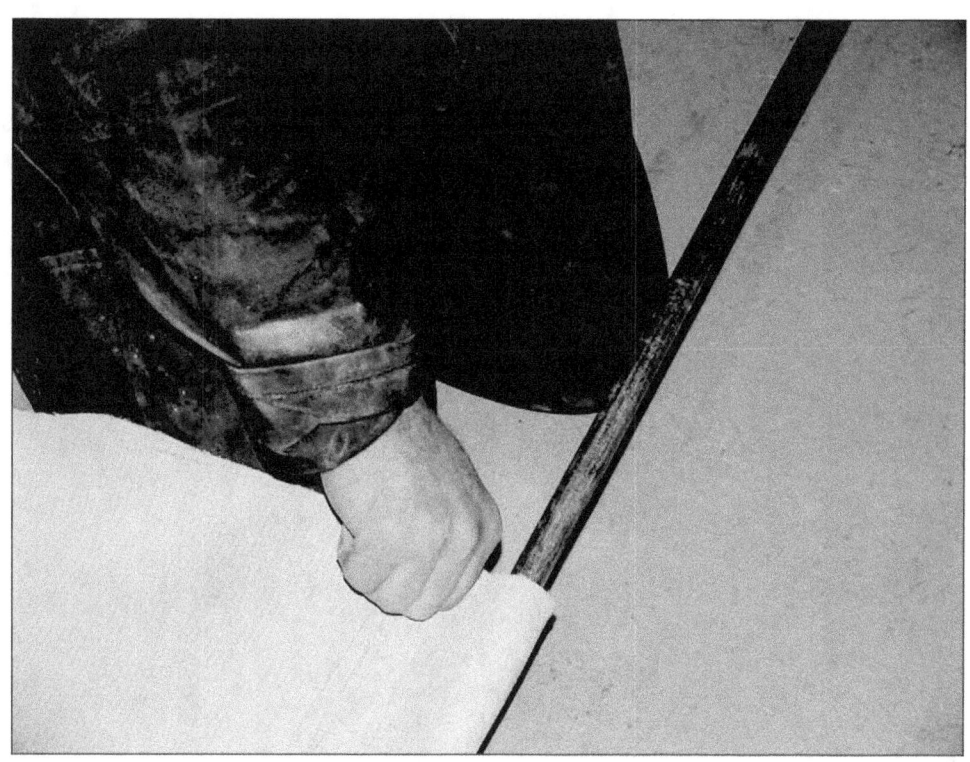

Figure 25.—Inserting pipe into top sleeve of woven cloth.

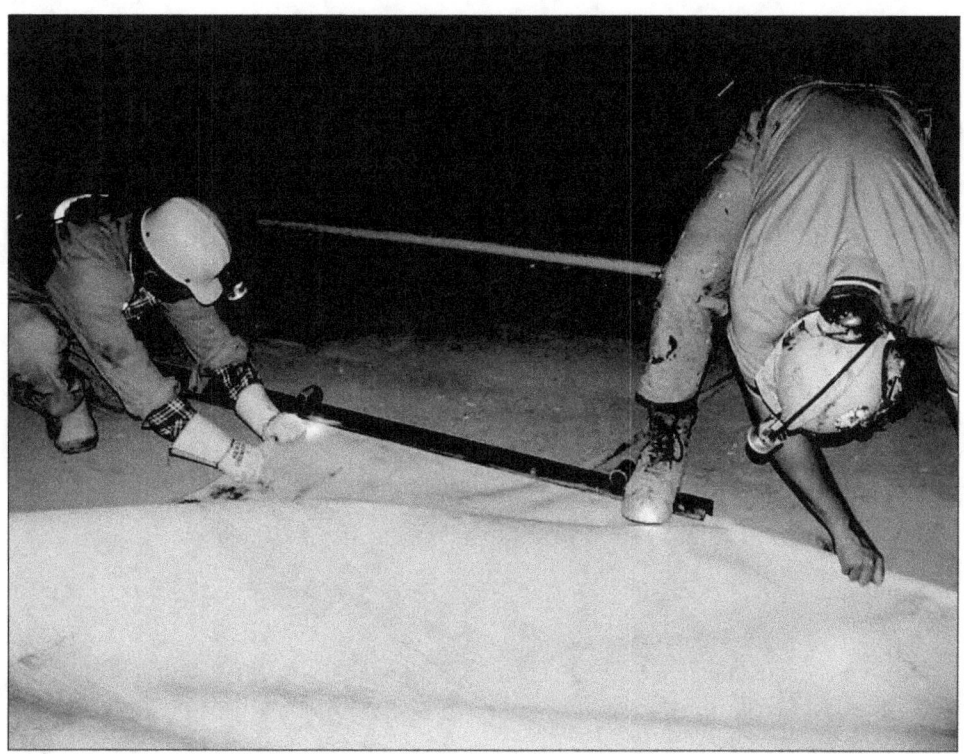

Figure 26.—Inserting pipe/sleeve into top box section.

Figure 27.—Anchoring top box section to the mine roof.

Figure 28.—Wedging woven cloth into mine rib.

Figure 29.—Sealant along perimeter of the stopping, viewed from C-drift.

Figure 30.—Rib bolts installed in the X-6 stopping, viewed from B-drift.

Figure 31.—Completed woven cloth stopping design in X-6, viewed from the explosion or C-drift side.

EXPLOSION TEST RESULTS

Dry-Stacked Hollow-Core Concrete Block Stoppings in Crosscuts

The four dry-stacked hollow-core concrete block stoppings were evaluated in a series of LLEM explosion tests (#427–#430 and #432–#434). A summary of these LLEM explosion tests is presented in Table 1. The first two columns in the table list the test number and date. The third column describes the ignition zone, where "C0-10 gas" means a methane-air zone in C-drift from 0 to 10 ft from the face. The peak explosion pressure generally occurs near the face of C-drift; the pressures are lower at the positions of the stoppings further from the face. A complete listing of the B- and C-drift omnidirectional wall pressures and the stopping pressures for each test is given in Tables A-1 through A-7 in the appendix. The stopping pressure is the total explosion pressure, which is also the omnidirectional pressure until such time as the stopping would fail. Note that the pressure results were averaged over 10 ms (15-point smoothing). Some of these tests were conducted for purposes other than the explosion evaluation of the stoppings. This is why some tests had lower explosion pressures than a preceding test.

Table 1.—Evaluations of mine stoppings during the LLEM explosion tests

Test #	Date	Ignition zone	Match location	Peak explosion wall pressure[4] psi	kPa	Result
427	Feb. 12, 2003	C0-10 gas	C10	1.2	9	No damage.
428	Feb. 13, 2003	C0-10 gas	Face	5.1	35	X-4 and X-5 hollow block stoppings destroyed.
429	Feb. 26, 2003	C0-10 gas[1]	C10[3]	3.7	26	Little damage to X-6 and X-7 hollow block stoppings.
430	Mar. 6, 2003	C0-10 gas[1]	Face[3]	6.2	43	Some block damage to X-6 hollow block stopping.
432	Mar. 13, 2003	C0-10 gas[1]	C10[3]	3.6	25	Additional block damage to X-6 hollow block stopping.
433	Mar. 17, 2003	C0-10 gas[1]	Face[3]	8.9	61	X-6 hollow block stopping destroyed.
434	Mar. 25, 2003	C0-10 gas[2]	Face[3]	7.6	52	X-7 hollow block stopping destroyed.
457	Oct. 27, 2003	C0-10 gas	C10	1.6	11	No damage.
458	Oct. 28, 2003	C0-10 gas	C6	1.9	13	X-6 and X-7 steel panel stoppings failed.
459	Nov. 13, 2003	C0-10 gas	C3	4.3	30	Some cracking on X-4 and X-5 solid block stoppings.
460	Nov. 18, 2003	C0-10 gas	Face[3]	6.3	43	Tearing of Australian stoppings; more cracking on solid block stoppings with some block displacement.
461	Nov. 20, 2003	C0-27 gas	Face	NA	NA	No additional damage.
462	Nov. 20, 2003	C0-27 gas	Face[3]	NA	NA	Solid block stoppings destroyed in X-4, X-6, and X-7; additional cracking/block displacement on X-5 solid block stopping.
463	Nov. 25, 2003	C0-47 gas	Face	~20	~138	X-5 solid block stopping destroyed.
491	May 10, 2005	B202-240 gas	B221	5.5	38	Hollow block stoppings destroyed.
494	June 9, 2005	B240-260 gas	B250	0.5	3	No damage.
495	June 10, 2005	B240-260 gas	B250	1.2	8	No damage.
496	June 13, 2005	B240-260 gas	B250	3.5	24	Hairline cracks on X-4 and B446 hollow block stoppings.
497	June 15, 2005	B202-240 gas	B221	5.1	35	X-4 and B446 hollow block stoppings destroyed; little to no damage to X-3 hollow block stoppings.
510	Dec. 11, 2007	A0-40 gas	Face	14.6	101	No damage.
512	Jan. 9, 2008	A0-40 gas A40-340 dust	Face	13.2	91	No damage.
515	Jan. 28, 2008	A0-50 gas	Face	14.0	96	No damage.
519	Mar. 3, 2008	A0-85 gas	Face	24.9	172	Hairline cracks on X-6 wet-laid solid block stopping; X-7 wet-laid solid block stopping destroyed.

NA Not available.
[1]Pulverized coal dust on shelving from C10-40 to result in a loading of 200 g/m^3.
[2]Pulverized coal dust on shelving from C10-40 (200 g/m^3 concentration) and a 200 g/m^3 coal loading with 65% added rock dust from C40-310.
[3]Ignition zone fan remained on during test to increase turbulence and peak overpressures.
[4]Peak omnidirectional pressure near the face; the stoppings were generally subjected to lower pressures.

Test #427

During test #427, there was little or no observable damage to the dry-stacked hollow-core concrete block stoppings, as listed in the last column of Table 1. The total explosion pressures at the stoppings in X-4 and X-7 were 0.74 psi (5.1 kPa) and 0.73 psi (5.0 kPa), respectively. The pressure transducers in front of the X-5 and X-6 stoppings did not work properly for this test, but the pressures there would have been similar to those at X-4 and X-7. The pressures in B-drift behind the stoppings were approximately zero (see Table A-1). There was no discernable movement on the LVDT sensor at the X-4 through X-6 stoppings, and the X-7 stopping had a permanent displacement of <0.05 in (<1.3 mm).

Test #428

During test #428, the dry-stacked hollow-core concrete block stoppings in X-4 and X-5 were destroyed by the explosion, as shown in Figure 32. The dry-stacked hollow-core concrete block stoppings in X-6 and X-7 showed little damage from this explosion. The left side of Figure 32 shows a plan view of the LLEM, with debris shown to the left of the original position of the stoppings in X-4 and X-5. The right side of the figure shows expanded cross-sectional views of the blocks remaining from the stoppings destroyed in X-4 and X-5. The LVDT displacement, the pressure-time integrals, and the pressures versus time at the stopping in X-4 and at nearby wall transducers during LLEM test #428 are shown in Figure 33. Graph A at the top of Figure 33 is the trace from an LVDT, which measures the displacement of the block near the center of the stopping. It shows that the stopping moved more than 3 in (8 cm) (maximum displacement that can be measured by the LVDT) as the stopping was destroyed. Graph B of Figure 33 shows the pressure-time integrals of the data from the inby (C-drift, 304 ft (93 m)) and outby (C-drift, 403 ft (123 m)) wall transducers, as well as the X-4C (355 ft (108 m)) transducer that is mounted from the mine roof at the midwidth of the crosscut just in front of the X-4 stopping location. Graph C of Figure 33 shows the total explosion pressure trace at the X-4 transducer, along with the pressures at the wall inby and outby the stopping location. The pressure pulse reaches the 304-ft (93-m) position first, then the stopping location, and finally the 403-ft (123-m) position. The maximum total explosion pressure from the X-4C transducer was 5.2 psi (36 kPa). The B-drift pressure behind the stopping was near zero until after the stopping was destroyed. Note that the total explosion pressure at the stopping was higher than the interpolated pressure (3.6 psi (25 kPa)) from the inby and outby wall pressure transducers. The inby and outby wall pressure transducers do not measure the total explosion pressures.

In previous literature [Weiss et al. 1999, 2002, 2006], the interpolated wall pressure (listed as the "static" pressure at the seal or stopping) was reported along with the "total pressure" at the seal or stopping. However, recent research has shown that this interpolated "static" wall pressure does not show the true explosion pressure exerted on the seal. The "total" pressure measured directly at the seal is higher than the interpolated wall pressure. Since the interpolated wall pressure is not the same as the actual explosion pressure at the seal or stopping location, it will no longer be used to describe the pressure at the seal or stopping. These different pressures are also discussed in the earlier "Experimental Mine and Test Procedures" section.

Figure 34 shows the debris from the destroyed X-4 stopping. Figure 35 shows a plan view map of the debris pattern, with black squares representing whole or partial blocks and the smaller markings representing smaller pieces of blocks. The original position of the stopping is shown by the double horizontal line in the crosscut near the bottom of the figure. The debris traveled into and beyond the intersection of X-4 and B-drift.

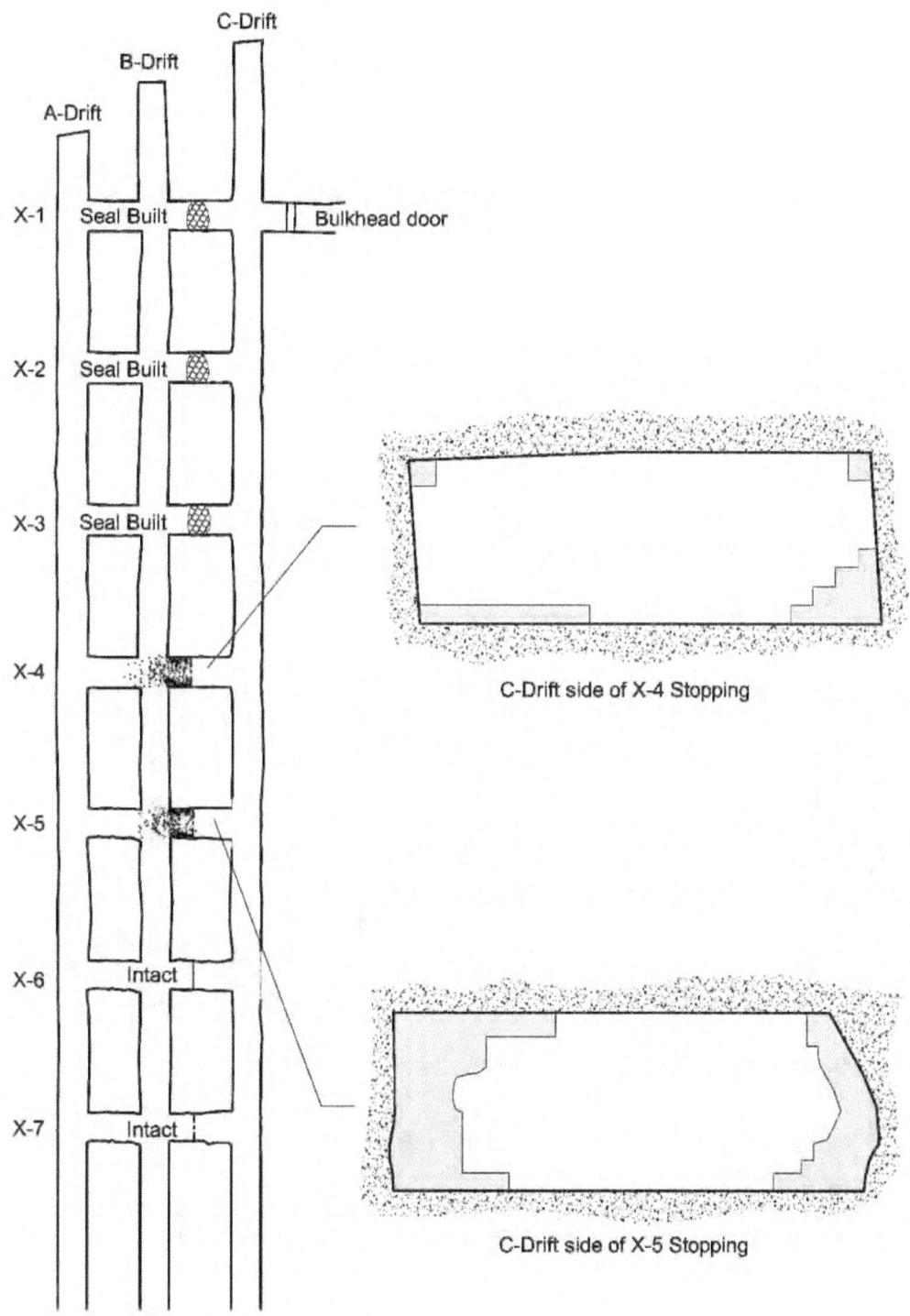

Figure 32.—Plan view of the multiple-entry area of LLEM showing effects of explosion test #428. The explosion destroyed the dry-stacked hollow-core concrete block stoppings in X-4 and X-5 and left the stoppings in X-6 and X-7 intact. The right part of the figure shows expanded cross-sectional views of the blocks remaining in X-4 and X-5, as seen from C-drift.

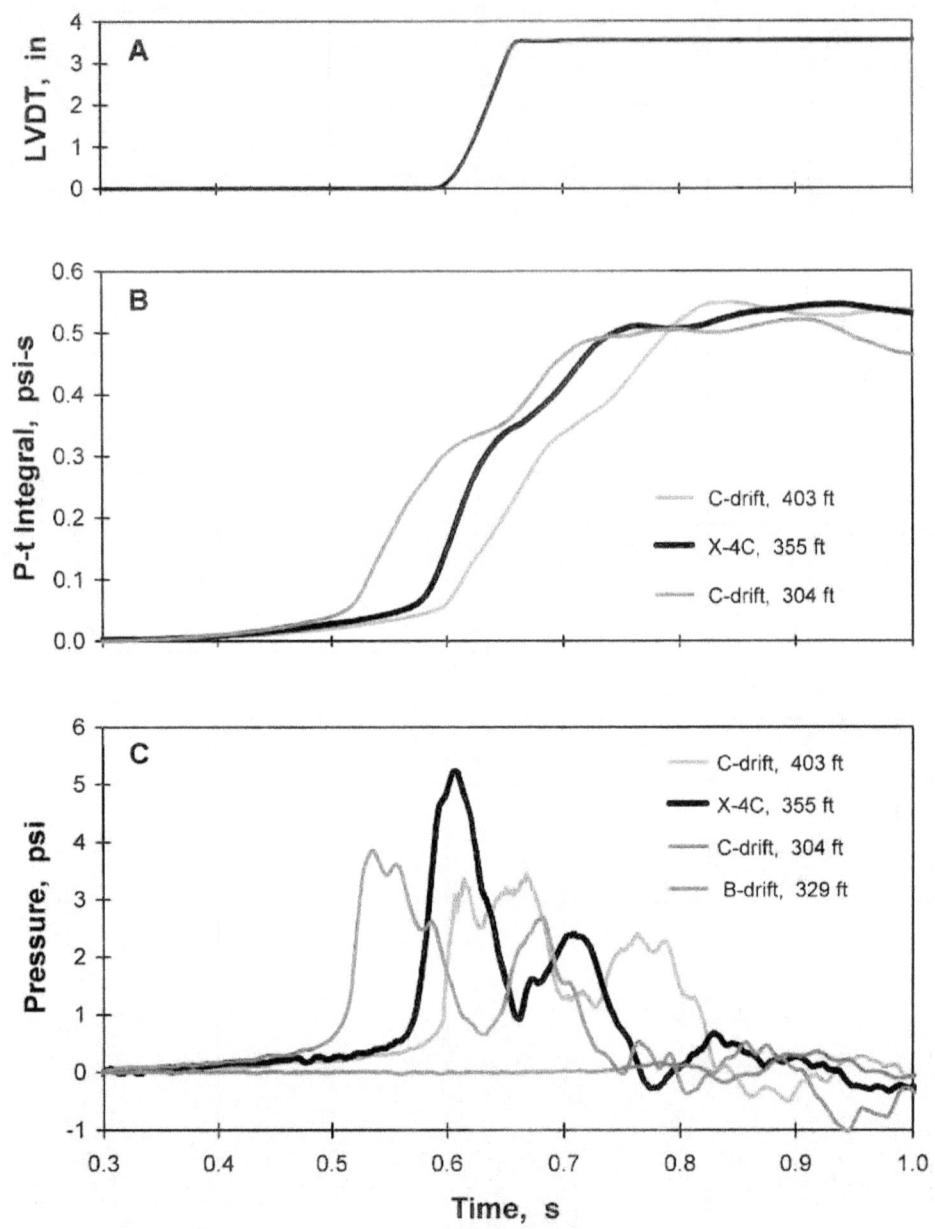

Figure 33.—(A) LVDT displacement, (B) pressure-time integrals, and (C) pressures versus time at dry-stacked hollow-core concrete block stopping in X-4 during LLEM test #428.

Figure 34.—Debris from dry-stacked hollow-core concrete block stopping in X-4 after LLEM test #428, viewed from C-drift.

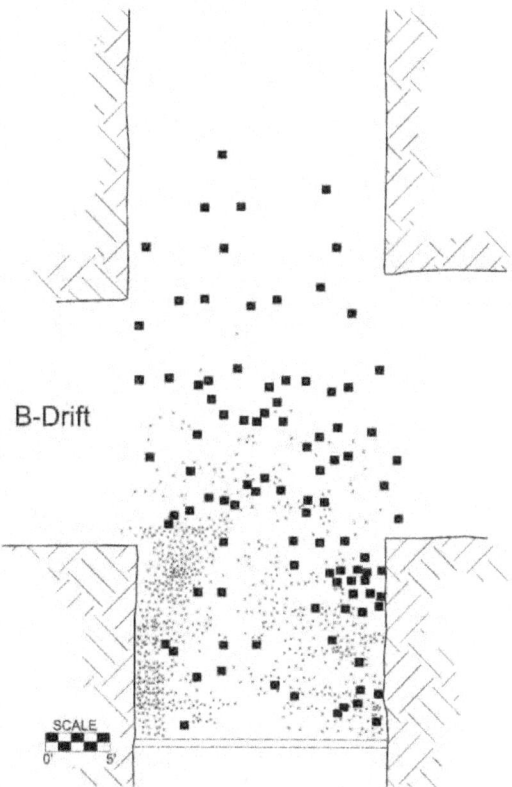

Figure 35.—Expanded map of debris from the X-4 dry-stacked hollow-core concrete block stopping after LLEM test #428. Black squares represent whole or partial blocks, and smaller markings represent pieces of blocks.

Figure 36 shows similar LVDT displacement, pressure-time integral, and pressure versus time data at and near the stopping in X-5, which was also destroyed during explosion test #428. There was no total explosion pressure measurement at X-5, because that transducer did not work properly during the test. The interpolated omnidirectional pressure from the wall-mounted DG panel sensors inby and outby the stopping location was 3.4 psi (23 kPa). The total explosion pressure at the X-5 stopping likely would have been higher, based on the data from X-4. The LVDT data showed that the X-5 stopping moved more than 3 in (8 cm) as the stopping was destroyed. Figure 37 shows the debris from the destroyed stopping in X-5. Figure 38 shows a plan view map of the debris from the X-5 stopping, with black squares representing whole or partial blocks and the smaller markings representing smaller pieces of blocks. The debris traveled into the intersection of X-5 and B-drift.

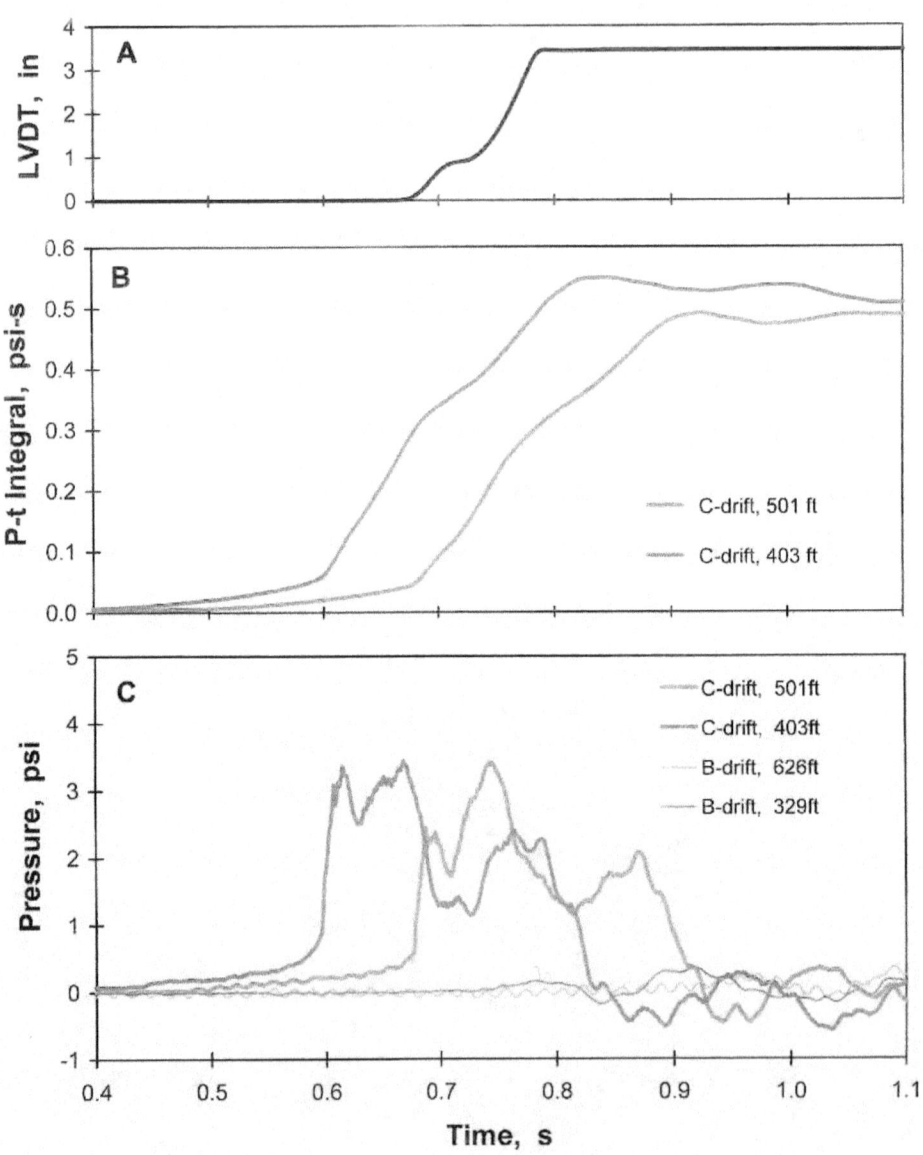

Figure 36.—(A) LVDT displacement, (B) pressure-time integrals, and (C) pressures versus time at the dry-stacked hollow-core concrete block stopping in X-5 during LLEM test #428.

Figure 37.—Debris from X-5 dry-stacked hollow-core concrete block stopping after LLEM test #428, viewed from C-drift.

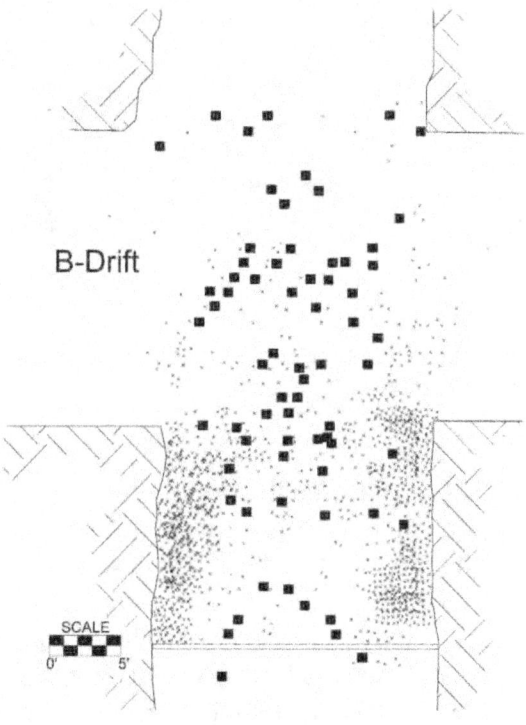

Figure 38.—Expanded map of debris from the X-5 dry-stacked hollow-core concrete block stopping after LLEM test #428. Black squares represent whole or partial blocks, and smaller markings represent pieces of blocks.

Remnants of the X-4 stopping were still attached to much of the floor, and a few blocks remained at the roof/rib interfaces. Mortar used at the floor to level the first course of block may have increased the overall strength of the stoppings. For the X-5 stopping, large sections of the original stopping were still attached along the entire interface of both ribs (Figure 32). There was clear evidence of block shearing along the edges of these still intact areas of the stoppings, indicating failure was mainly due to excessive forces acting on the whole wall. Figure 39 shows the inby crosscut rib remnants (viewed from B-drift) of the X-5 dry-stacked hollow-core concrete block stopping where block shearing occurred. This block shearing is a clear indication of roof-to-floor arching action.

Figure 39.—Remnants of the dry-stacked hollow-core concrete block stopping in X-5 after LLEM test #428. The block shearing indicates roof-to-floor arching action.

Figure 40 shows the LVDT displacement, pressure-time integral, and pressure versus time data at and near the dry-stacked hollow-core concrete block stopping in X-6 that withstood the explosion pressures generated during test #428. There was no total explosion pressure measurement at X-6 because that transducer did not work properly during the test. There was little damage to the X-6 stopping, as shown in Figure 41. The interpolated omnidirectional pressure from the wall-mounted DG panel sensors inby and outby the stopping in X-6 as shown in graph C of Figure 40 was 3.2 psi (22 kPa). The total explosion pressure at the X-6 stopping likely would have been higher, based on the data from X-4 (Figure 33). The LVDT data (Figure 40) showed almost no movement of the stopping.

Similarly, Figure 42 shows the LVDT displacement, pressure-time integral, and pressure versus time data at and near the dry-stacked hollow-core concrete block stopping in X-7. This stopping survived the test #428 explosion, and the LVDT showed a movement of less than 0.2 in (5 mm). The total explosion pressure was 3.4 psi (23 kPa). The interpolated omnidirectional pressure from the wall-mounted DG panel sensors inby and outby the stopping was lower (2.9 psi (20 kPa)).

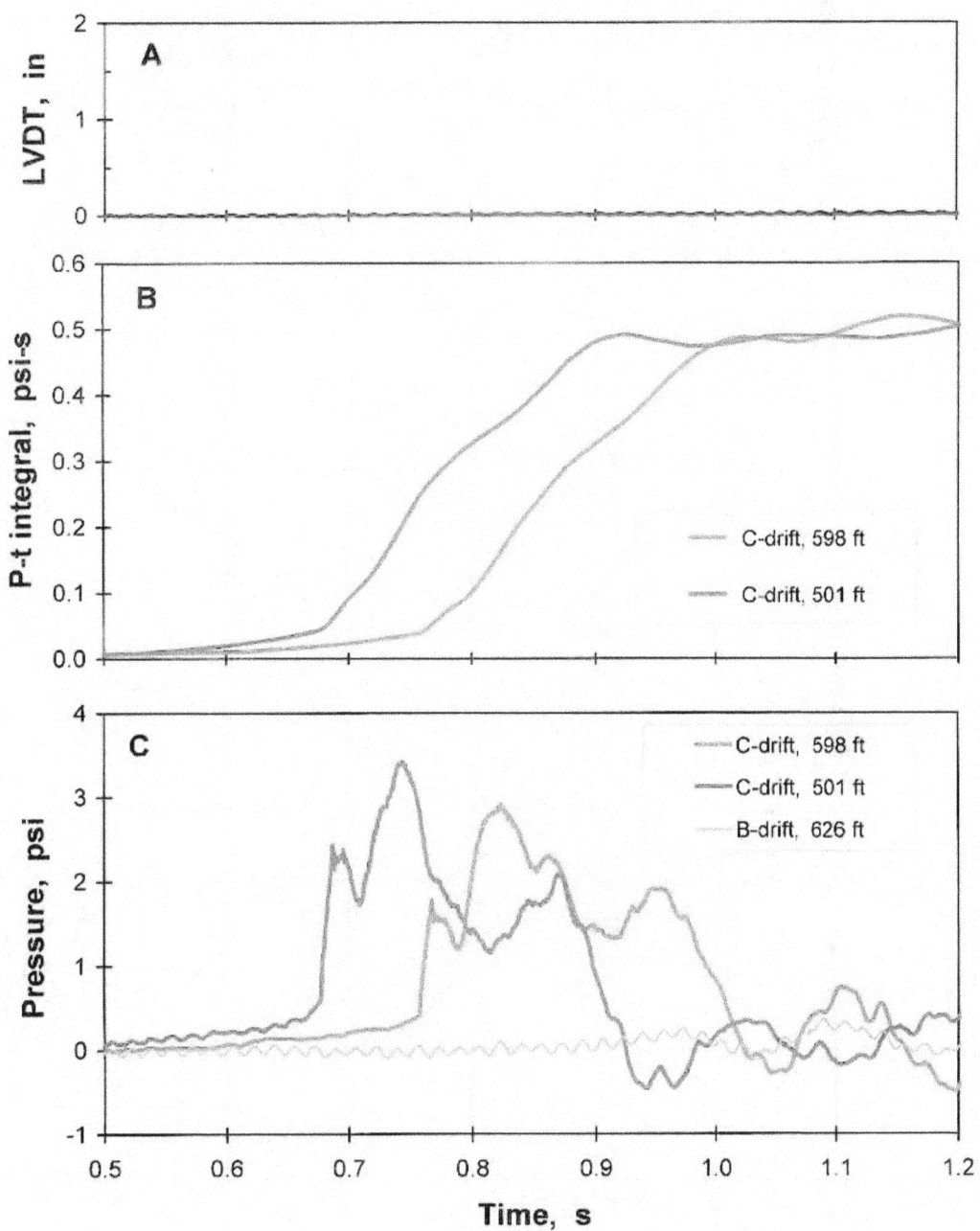

Figure 40.—(A) LVDT displacement, (B) pressure-time integrals, and (C) pressures versus time at the dry-stacked hollow-core concrete block stopping in X-6 during LLEM test #428.

Figure 41.—Dry-stacked hollow-core concrete block stopping in X-6 remained intact after LLEM test #428. Chalk highlighted the cracks in the surface coating material.

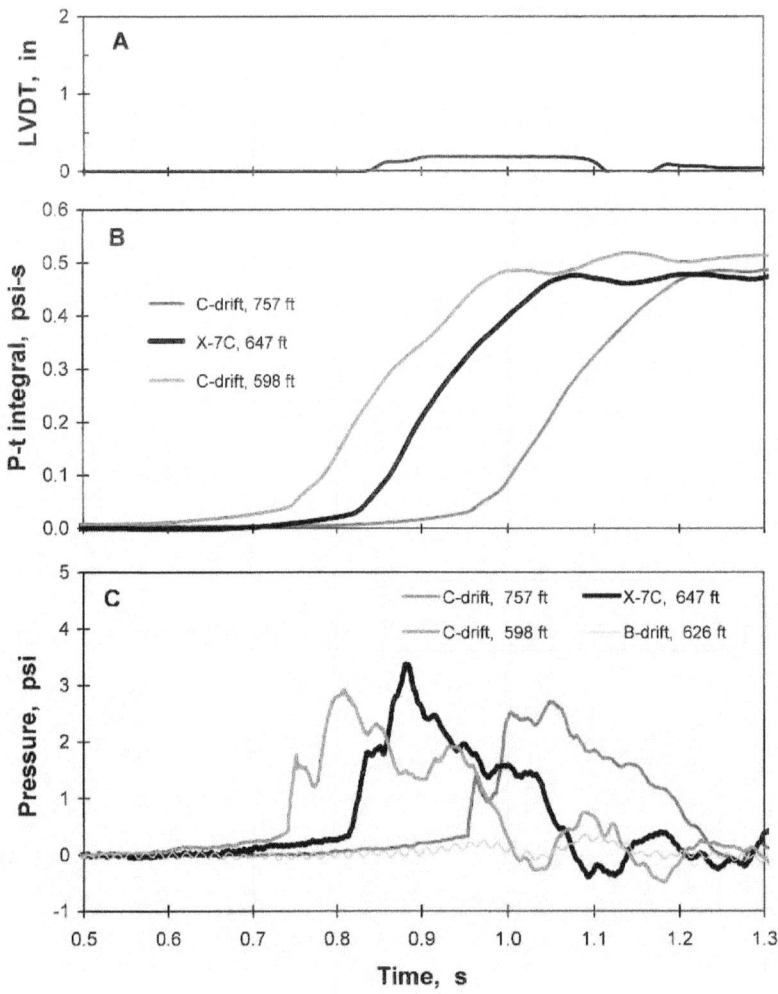

Figure 42.—(A) LVDT displacement, (B) pressure-time integrals, and (C) pressures versus time at the dry-stacked hollow-core concrete block stopping in X-7 during LLEM test #428.

A summary of the pressure data at the dry-stacked hollow-core concrete block stopping positions is shown in Figure 43 and Table 2. The peak omnidirectional wall pressures in C-drift at the various DG panel positions are shown in Figure 43 as a function of distance from the face. Note that the omnidirectional wall pressures decrease from about 5 psi (35 kPa) near the face to slightly less than 3 psi (21 kPa) at 757 ft (231 m) from the face. At the top of Figure 43 is a schematic showing the locations of the four stoppings in X-4, X-5, X-6, and X-7 at distances of 355, 451, 547, and 647 ft (108, 138, 167, and 197 m) from the face, respectively. The B-drift pressures are the maximum values up until 0.1 sec after peak pressure in C-drift at the same distance from the face. (The pressures in B-drift were somewhat higher at later times due to the pressures from the destroyed stoppings.) The total explosion pressures were only available at the X-4 and X-7 stoppings. The pressure-time integrals in Table 2 are the average values up until the pressure first goes negative; these do not include any contributions from later secondary peaks (Figures 33 and 42). Because the pressure transducers at the stoppings in X-5 and X-6 did not work properly during the test, there are no data for these stoppings in Table 2. In summary, this type of dry-stacked hollow-core concrete block stopping, as constructed in the LLEM, survived a total explosion pressure of 3.4 psi (23 kPa) and was destroyed by a total explosion pressure of 5.2 psi (36 kPa) during this evaluation. If total explosion pressure data had been available for the X-5 and X-6 stoppings, the known range of pressures from survival to destruction would probably have been better defined.

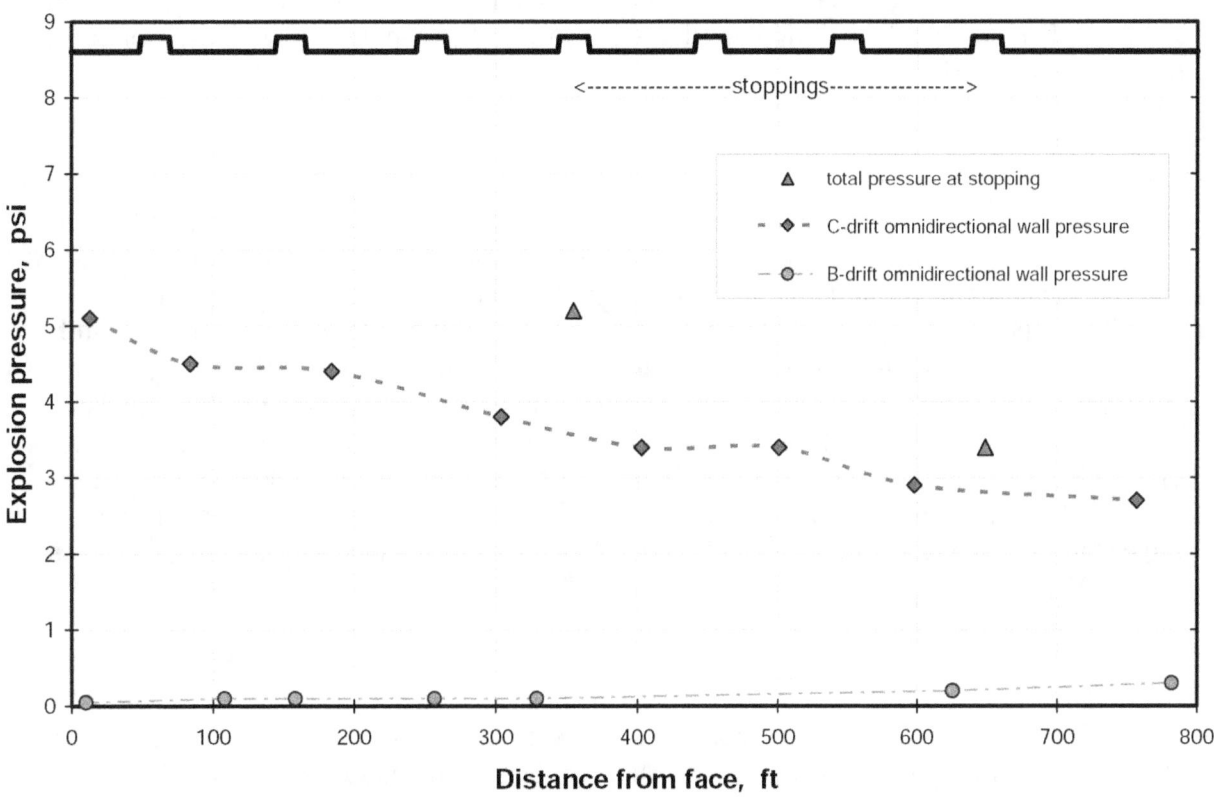

Figure 43.—Pressures at walls (ribs) and at dry-stacked hollow-core concrete block stopping locations during LLEM test #428.

Table 2.—Peak total explosion pressure data at dry-stacked hollow-core concrete block stoppings for test #428 in C-drift

Location	Distance		Peak total explosion pressure		Pressure-time integral	
	ft	m	psi	kPa	psi-s	kPa-s
Crosscut 4	355	108	5.2	36	0.51	3.5
Crosscut 5	451	138	NA	NA	NA	NA
Crosscut 6	547	167	NA	NA	NA	NA
Crosscut 7	647	197	3.4	23	0.48	3.3

NA Not available (pressure transducers did not work properly during the test).

Test #429

During test #429, the explosion pressures were lower. All of the pressure transducers at the stopping locations worked properly for this test. The total explosion pressure was ~1.2 psi (~8 kPa) at the X-6 dry-stacked hollow-core concrete block stopping. The total explosion pressure at the X-7 stopping was ~1.0 psi (~7 kPa). The LVDTs showed movements of less than 0.2 in (4 mm). There was little observable damage to these two remaining stoppings during test #429.

The LVDT data for the X-6 and X-7 stoppings were questionable. The signals may have been associated with the explosion pressures coming through the open inby crosscuts (X-4 and X-5) and then moving the fishing lines that were strung between the X-6 and X-7 stoppings and the sensors in B-drift.

Test #430

Although the pressure near the face was higher for test #430 than for test #428 (Table 1), the pressures at the stoppings were not any higher because of the venting through the open X-4 and X-5. The total explosion pressure was 3.8 psi (26 kPa) at the X-6 stopping and 2.2 psi (15 kPa) at the X-7 stopping. During this explosion, about three blocks were knocked out from the X-6 stopping and other blocks were damaged. Expanded drawings of the damage to the X-6 stopping, as viewed from both sides, are shown in Figure 44. Figures 45–46 show the damage to the X-6 stopping, as viewed from B-drift. Figure 47 shows the damage to the X-6 stopping, as viewed from C-drift. The damage to the X-6 stopping was due mainly to the shearing of the block on the center sections of the first and second block courses, which resulted from the arching of the stopping wall. As shown in Figure 48, there was little observable damage to the stopping in X-7.

Test #432

During test #432, the explosion pressures were lower than those for test #430. The total explosion pressure was ~1.7 psi (~12 kPa) at the X-6 stopping and ~1.2 psi (~9 kPa) at the X-7 stopping. During this explosion, an additional couple of blocks were knocked out from the X-6 stopping. There was little observable damage to the X-7 stopping. By this time, the two remaining dry-stacked hollow-core concrete block stoppings in X-6 and X-7 had been subjected to multiple explosions and, to some extent, had been damaged and/or weakened.

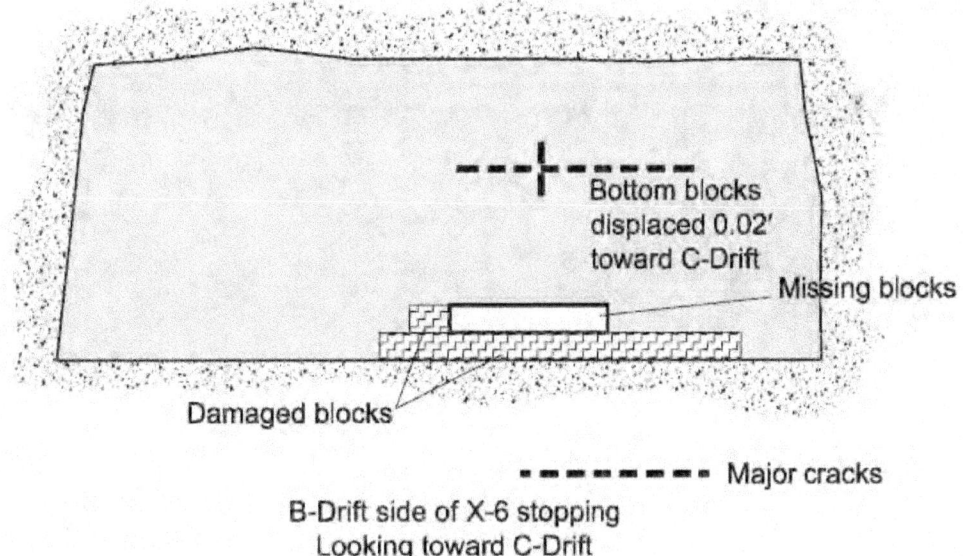

B-Drift side of X-6 stopping
Looking toward C-Drift

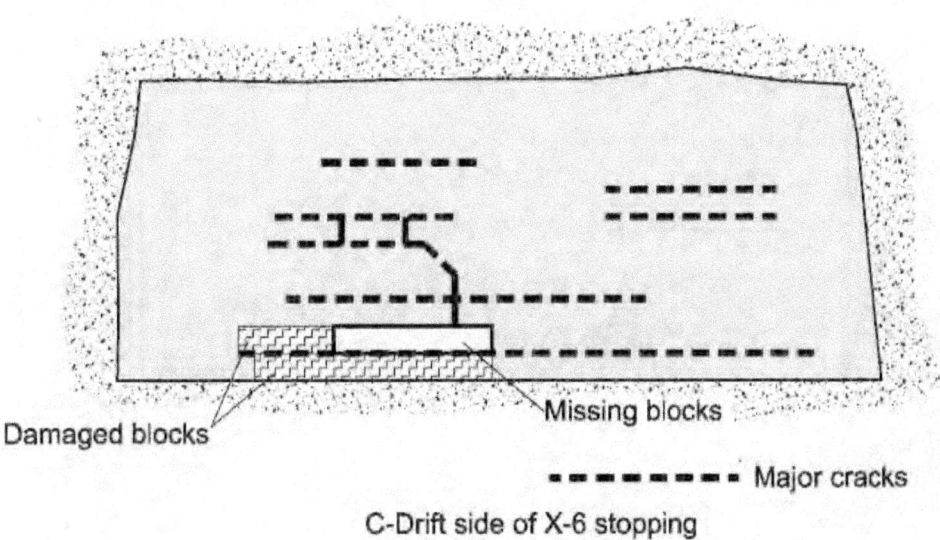

C-Drift side of X-6 stopping

Figure 44.—Expanded cross-sectional views of the blocks remaining in X-6 after test #430.

Figure 45.—Crosscut 6 dry-stacked hollow-core concrete block stopping viewed from B-drift side after LLEM test #430, showing blocks dislodged by explosion.

Figure 46.—Closeup of B-drift side of X-6 dry-stacked hollow-core concrete block stopping after test #430, showing dislodged and damaged blocks. This is clear evidence of roof-to-floor arching action.

Figure 47.—Closeup of C-drift side X-6 dry-stacked hollow-core concrete block stopping after test #430, showing missing and damaged blocks.

Figure 48.—The dry-stacked hollow-core concrete block stopping in X-7 remained intact after LLEM test #430.

Test #433

During test #433, the total explosion pressure was ~3.6 psi (~25 kPa) at the X-6 stopping, and this stopping was mostly destroyed (see Figures 49–51). Only about 5–6 ft (1.5–1.8 m) of the inby part of the X-6 stopping remained after the explosion, as shown in Figures 49 and 51. Also shown in Figures 50–51 is the pressure transducer that was suspended from the roof in front of the stopping location. A plan view map of the debris from the X-6 stopping is shown in Figure 52. At the X-7 stopping, the total explosion pressure was ~2.8 psi (~19 kPa). There was one block that was partially dislodged in the X-7 stopping, and several other blocks had broken faces. Figure 53 shows the damage to the C-drift side of the X-7 stopping. Figure 54 shows an overview of the C-drift side of the X-7 stopping after test #433. There is some damage to the blocks at the top of the stopping just to the right of the pressure transducer. This is shown in more detail in Figure 55. Figure 56 shows one block that was almost dislodged from the X-7 stopping, as noted in the Figure 53 drawing. This block movement is attributed mainly to a failure of the coating.

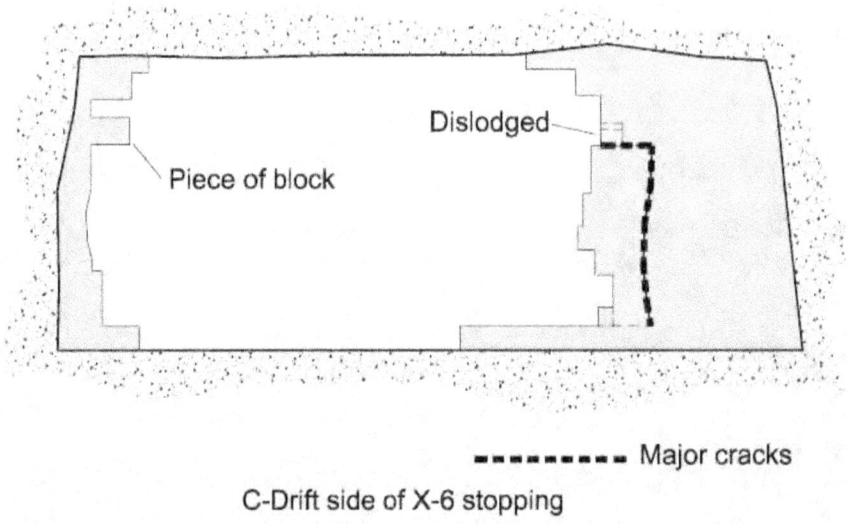

Figure 49.—Expanded cross-sectional view of the blocks remaining in X-6 after test #433.

Figure 50.—Debris from X-6 dry-stacked hollow-core concrete block stopping after LLEM test #433, viewed from C-drift.

Figure 51.—Part of the inby side of X-6 dry-stacked hollow-core concrete block stopping remaining after LLEM test #433, viewed from C-drift.

B-Drift

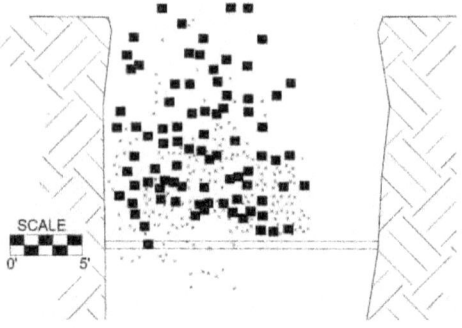

Figure 52.—Expanded map of debris from the X-6 dry-stacked hollow-core concrete block stopping after LLEM test #433. Black squares represent whole or partial blocks, and smaller markings represent pieces of blocks.

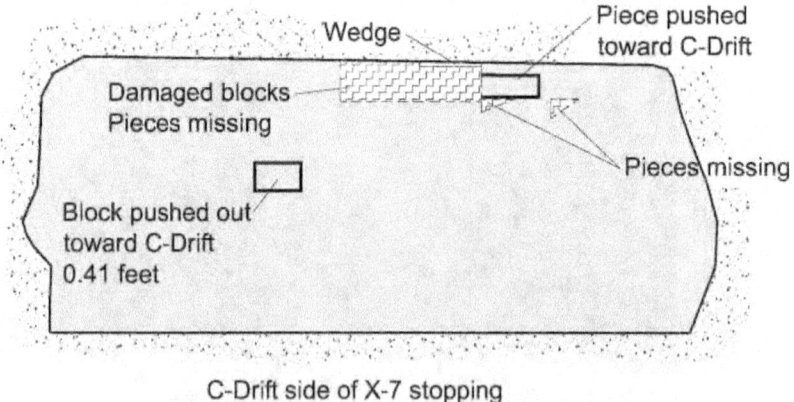

C-Drift side of X-7 stopping

Figure 53.—Expanded cross-sectional view of the blocks remaining in X-7 after explosion test #433, as seen from C-drift side.

Figure 54.—Condition of X-7 dry-stacked hollow-core concrete block stopping after explosion test #433, viewed from C-drift side.

Figure 55.—Expanded view of top part of X-7 dry-stacked hollow-core concrete block stopping after explosion test #433, showing damaged blocks.

Figure 56.—Expanded view of hollow-core concrete block that is almost dislodged from X-7 dry-stacked stopping after explosion test #433.

Test #434

All of the previous LLEM explosion tests (#429–#430, #432–#433) had a small methane-air zone, plus a short zone of coal dust. Test #434 had a long zone (40–310 ft (12–94 m)) of 35% bituminous coal dust and 65% limestone rock dust, and the explosion flame traveled to ~770 ft (~235 m). Although the total explosion pressure at the one remaining dry-stacked hollow-core concrete block stopping in X-7 was only ~2.3 psi (~16 kPa), a large section of the X-7 stopping was knocked out during the explosion. Damage from previous explosion tests likely weakened the X-7 stopping, resulting in failure at a lower-than-expected total explosion pressure. The pressure-time integrals are much larger for the dust explosion test (3.0 psi-s (21 kPa-s)) compared to the previous gas explosion tests (0.48-0.66 psi-s (3.3–4.6 kPa-s)) at this stopping location. Figure 57 shows the plan view of the LLEM multiple-entry area and cross-sectional drawings of the remaining blocks left from the stoppings in X-6 and X-7. Figure 58 shows additional blocks partially dislodged from the X-6 stopping. Figures 59 and 60 show the B- and C-drift sides, respectively, of the X-7 stopping after test #434. There was much less debris on the C-drift side of the stopping compared to the B-drift side, and the C-drift debris was located very near the stopping. Figure 61 shows a map of the debris from the X-7 stopping. Figures 59–61 show there was debris from the X-7 stopping on both the B- and C-drift sides.

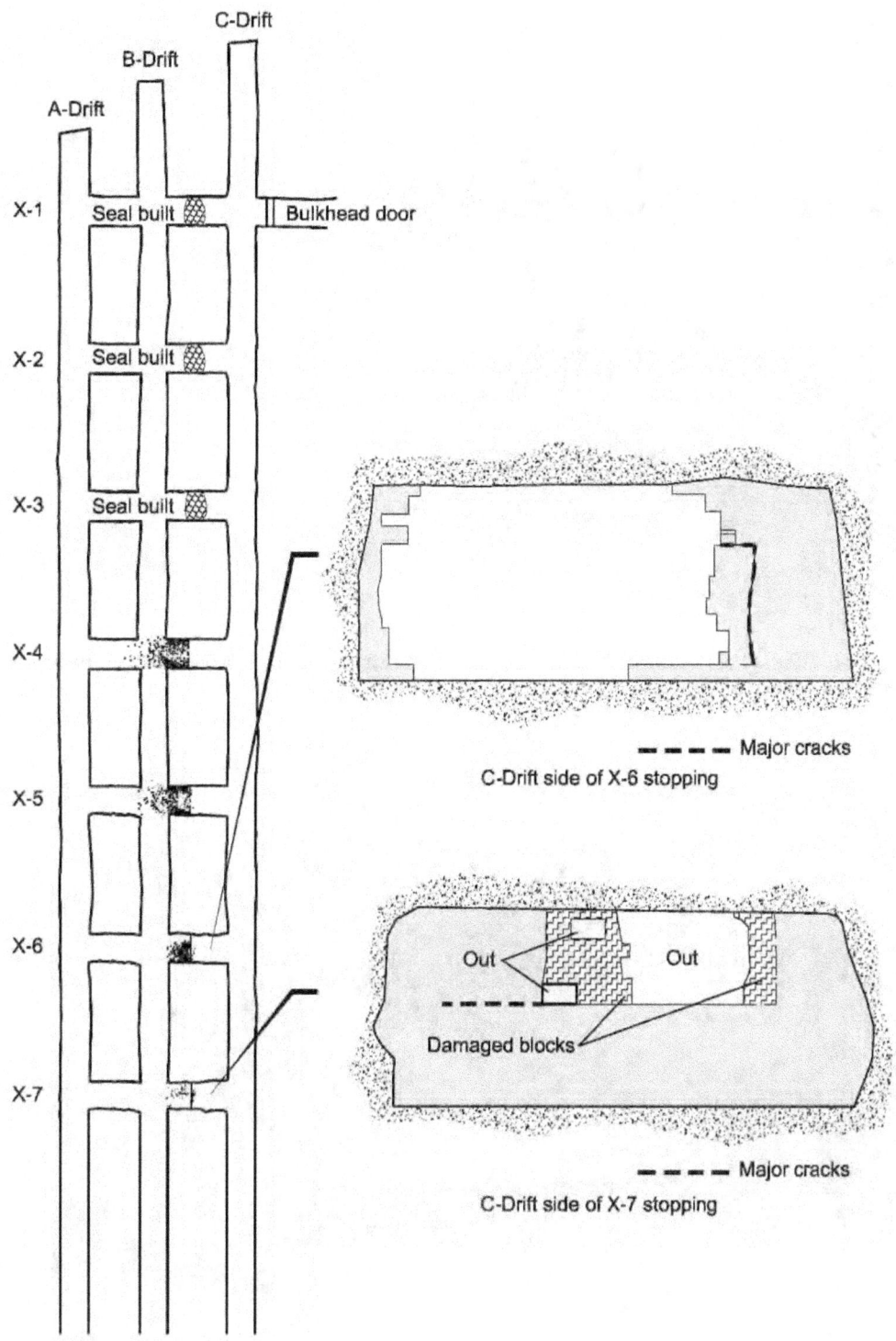

Figure 57.—Effects of explosion test #434 on the dry-stacked hollow-core concrete block stoppings, with expanded cross-sectional views of the blocks remaining in X-6 and X-7.

Figure 58.—Additional blocks partially dislodged from the right side of X-6 dry-stacked hollow-core concrete block stopping after test #434, viewed from C-drift.

Figure 59.—Crosscut 7 dry-stacked hollow-core concrete block stopping viewed from B-drift after LLEM test #434.

Figure 60.—Crosscut 7 dry-stacked hollow-core concrete block stopping viewed from C-drift after LLEM test #434.

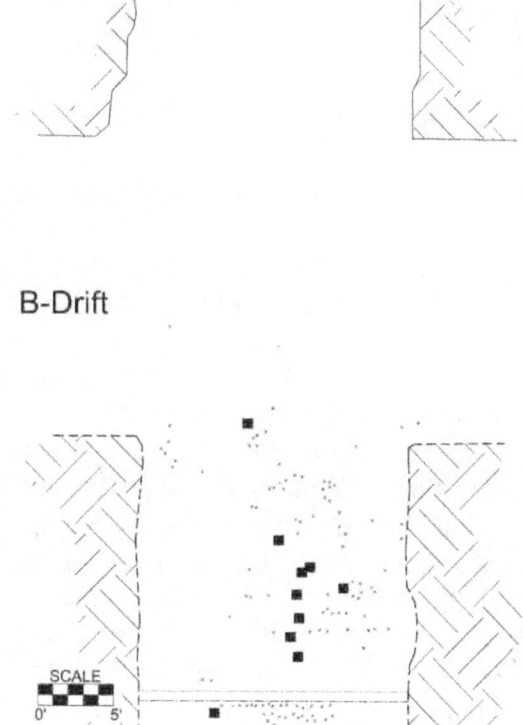

Figure 61.—Map of debris from the X-7 dry-stacked hollow-core concrete block stopping after LLEM test #434. Black squares represent whole or partial blocks, and smaller markings represent pieces of blocks.

Steel Panel Stoppings in Crosscuts

The two steel panel stoppings in X-6 and X-7 were evaluated during a series of two LLEM explosion tests (#457–#458). A summary of these LLEM explosion tests is presented in Table 1. A complete listing of the B- and C-drift omnidirectional wall pressures and the stopping pressures for each test is given in Tables A-8 and A-9 in the appendix.

Test #457

During test #457, there was little or no observable damage to either of the steel panel stoppings. The total explosion pressure was ~0.8 psi (~5.6 kPa) at the X-6 steel panel stopping and ~0.8 psi (~5.2 kPa) at the X-7 stopping. The pressures in B-drift behind the stoppings were negligible. The LVDT data showed ~0.4-in (~1-cm) maximum displacement for each of the two steel panel stoppings, but the stoppings returned close to their original positions after the explosion.

Test #458

During test #458, the steel panel stoppings in X-6 and X-7 failed during the explosion. The LVDT displacement, pressure-time integrals, and pressures versus time associated with the steel panel stopping in X-6 during test #458 are shown in Figure 62. The total explosion pressure measured with the transducer in front of the steel panel stopping is listed as "X-6C." The total explosion pressure measured at X-6 was 1.3 psi (9 kPa). The B-drift pressure behind the stopping in X-6 was near zero until after the stopping failed. At the top of Figure 62 in graph A is the trace from an LVDT showing that the stopping moved over 3 in (8 cm) and failed. Graph B of Figure 62 shows the pressure-time integrals at the position of the X-6 stopping. The reported pressure-time integral data in Table A-9 are the values up until the pressure first goes negative; they do not include any contributions from later secondary peaks. The debris from the failed X-6 steel panel stopping is shown in Figures 63–64. The three horizontal steel angles that were inset into the X-6 ribs on the B-drift side of the stopping deformed and pulled out of the shallow rib insets (2 in (5 cm) deep) as the C-drift explosion pressure was exerted on the stopping. This deformation of the steel angles can be seen in Figure 63. The stopping in X-6 was displaced from its original position as a unit, with only a few detached panels.

Similar pressure data were collected for the steel panel stopping in X-7. Figure 65 shows the LVDT displacement, pressure-time integral, and pressure versus time data for this stopping. The total explosion pressure at the stopping in X-7 as shown in graph C of Figure 65 was 1.3 psi (9 kPa). The LVDT also showed over 3 in (8 cm) of movement as the stopping failed (graph A of Figure 65). Figures 66–67 show the postexplosion condition of the X-7 steel panel stopping. After the test, one side of the X-7 stopping was still partially attached to the outby rib, but the remainder of the stopping was on the floor. The three horizontal steel angles that were inset into the X-7 inby rib deformed and pulled out of the rib insets as the explosion pressure was exerted on the stopping.

In summary, this particular type of steel panel stopping as constructed in the LLEM survived a total explosion pressure of ~0.8 psi (~5.5 kPa) and failed at a total explosion pressure of 1.3 psi (9 kPa). The primary mode of failure for this steel panel stopping may be more related to the strength and deformation of the horizontal steel angles behind the stoppings than the failure of the steel panels. The method of failure indicates that the horizontal steel angles contribute to the integrity of the stopping. If the horizontal steel angle had been stronger and/or inset deeper into the crosscut ribs, the steel panel stoppings may have resisted higher explosion overpressures.

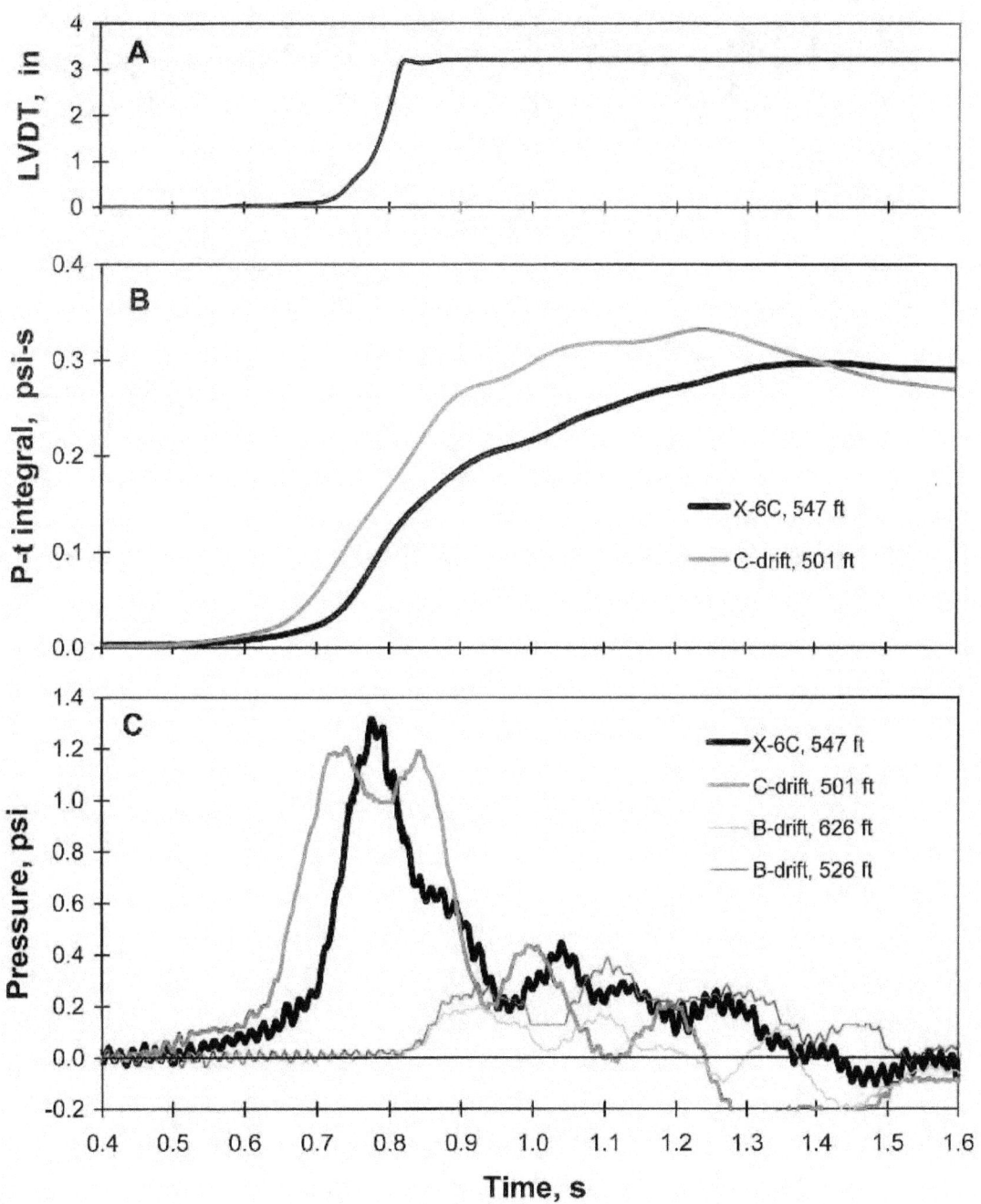

Figure 62.—(A) LVDT displacement, (B) pressure-time integrals, and (C) pressures versus time at the steel panel stopping in X-6 during LLEM test #458.

Figure 63.—Debris from steel panel stopping in X-6 after LLEM test #458, viewed from C-drift.

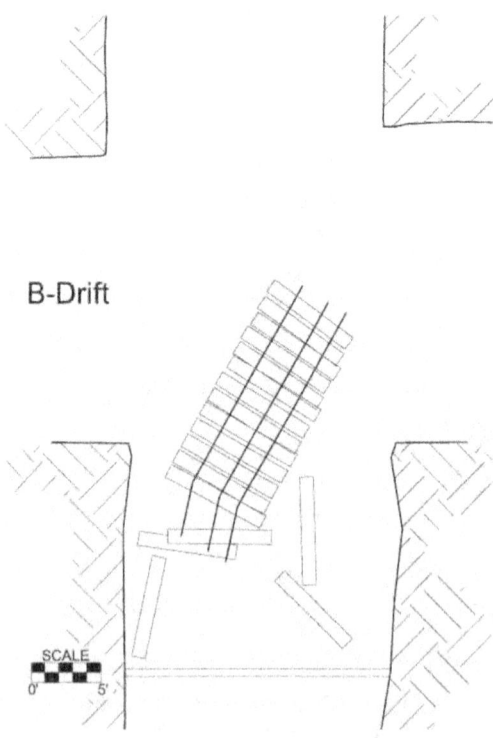

Figure 64.—Map of debris from the X-6 steel panel stopping after LLEM test #458.

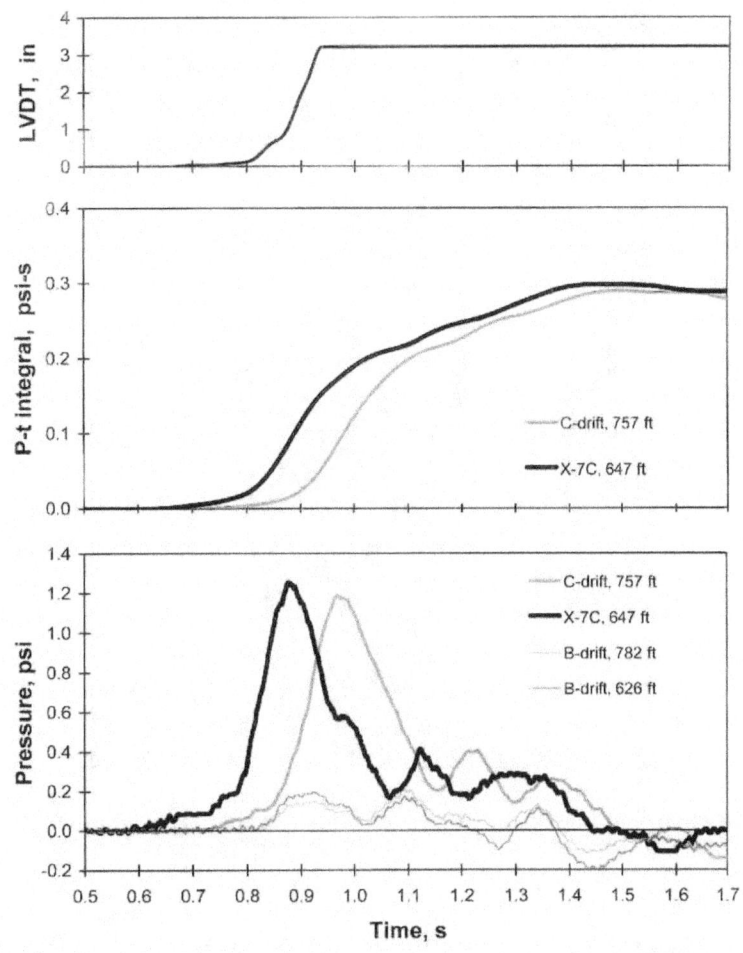

Figure 65.—(A) LVDT displacement, (B) pressure-time integrals, and (C) pressures versus time at the steel panel stopping in X-7 during LLEM test #458.

Figure 66.—Debris from steel panel stopping in X-7 after LLEM test #458, viewed from B-drift.

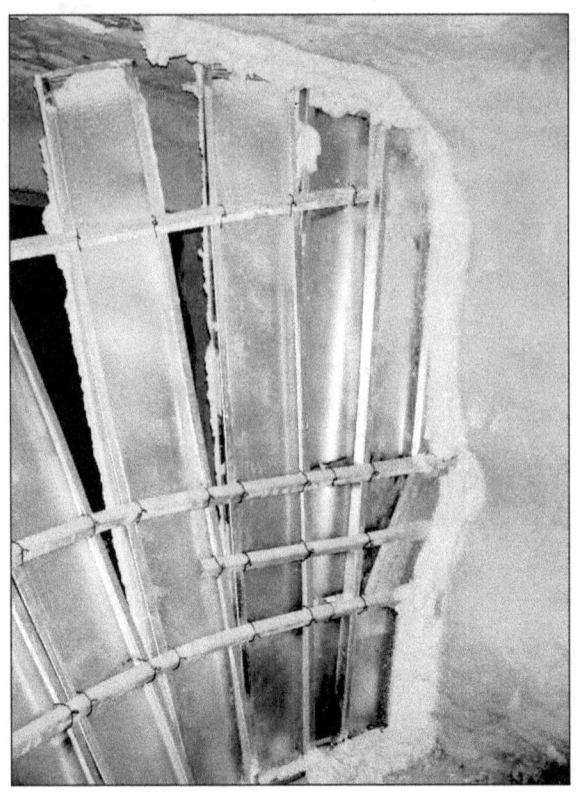

Figure 67.—Closeup of steel panel stopping still partially attached to outby rib in X-7 after LLEM test #458, viewed from B-drift.

Dry-Stacked Solid-Concrete-Block Stoppings in Crosscuts

The two dry-stacked solid-concrete-block stoppings in X-4 and X-5 were evaluated during a series of seven explosion tests (#457–#463). A summary of these LLEM explosion tests is presented in Table 1. A complete listing of the B- and C-drift omnidirectional wall pressures and stopping pressures for each test is given in Tables A-8 through A-14 in the appendix.

Test #457

During test #457, there was little or no observable damage to the two dry-stacked solid-concrete-block stoppings located in X-4 and X-5. The total explosion pressures were 0.78 psi (5.4 kPa) and 0.75 psi (5.2 kPa) at the stoppings in X-4 and X-5, respectively. The pressures in B-drift behind the stoppings were negligible (<0.1 psi (<1 kPa)). The LVDT showed no movement on the two stoppings.

Test #458

During test #458, the dry-stacked solid-concrete-block stoppings in X-4 and X-5 showed no noticeable damage. The LVDT displacement, pressure-time integrals, and pressure versus time traces associated with the dry-stacked solid-concrete-block stopping in X-4 are shown in Figure 68. The total explosion pressure measured with the transducer in front of the stopping is labeled as "X-4C" and shown as a solid black line. As shown in graph C of Figure 68, the solid-concrete-block stopping in X-4 was subjected to a total explosion pressure of 1.6 psi (11 kPa). As is shown in graph A of Figure 68, no significant movement (only ~0.06 in (~1.5 mm)) was observed on the LVDT. Similar displacement and pressure data were measured at the X-5 stopping during this test. The stoppings returned close to their original positions after the explosion.

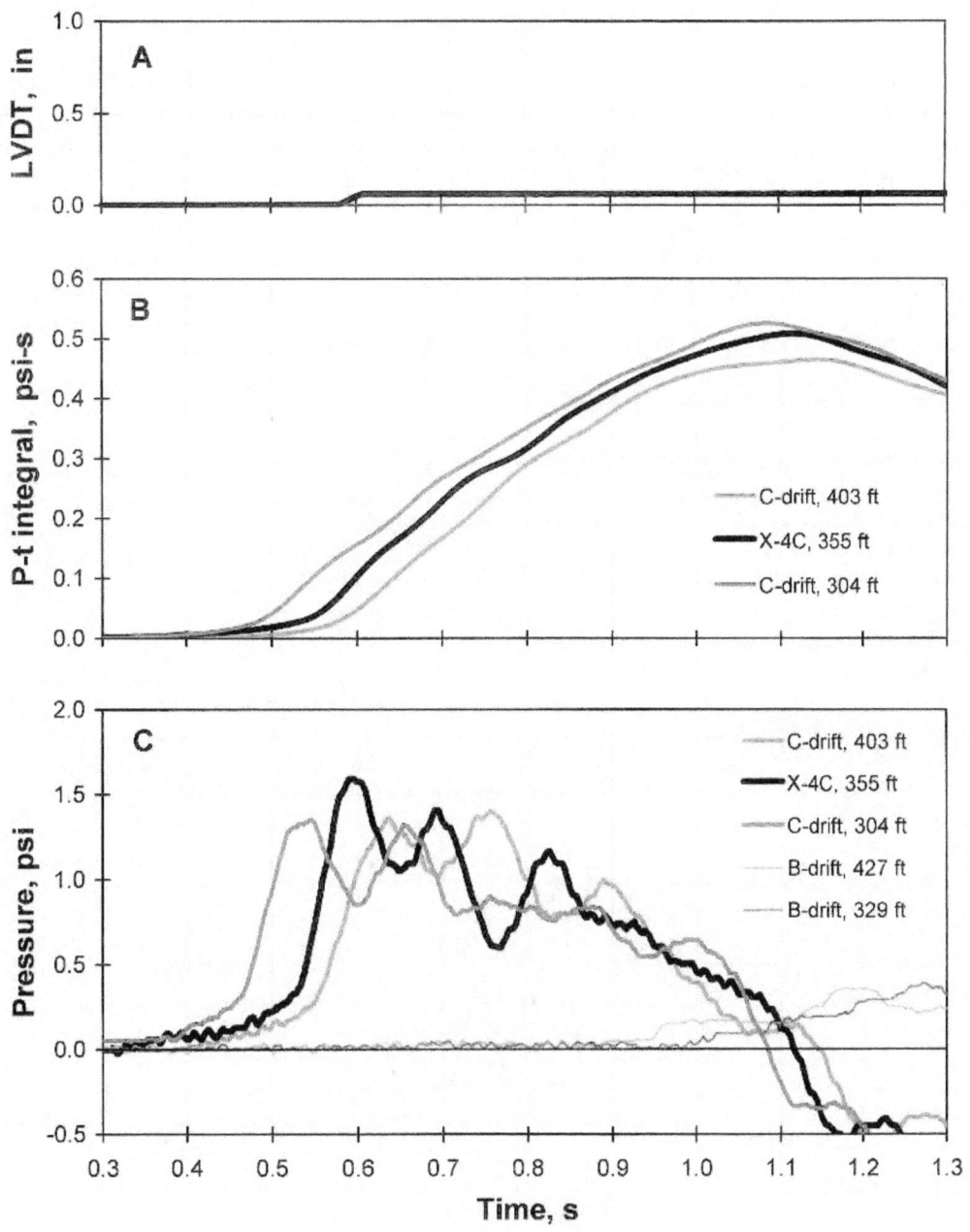

Figure 68.—(A) LVDT displacement, (B) pressure-time integrals, and (C) pressures versus time at the dry-stacked solid-concrete-block stopping in X-4 during LLEM test #458.

Test #459

During test #459 (see Table A-10), the total explosion pressure was 4.6 psi (32 kPa) at the X-4 dry-stacked solid-concrete-block stopping (graph C of Figure 69). For the X-5 stopping, the total explosion pressure was 3.4 psi (23 kPa). The LVDTs showed movement of nearly 0.7 in (18 mm) for the X-4 stopping (graph A of Figure 69) and 0.4 in (10 mm) for the X-5 stopping. After the explosion, the X-4 stopping had a permanent displacement of ~0.3 in (~8 mm) at the midpoint sensor location, and the X-5 stopping a permanent displacement of ~0.15 in (~4 mm) at the midpoint sensor location. However, both stoppings survived, with only limited damage. Both stoppings exhibited near rib-to-rib cracking of the sealant coating between the first and second block courses from the floor, with an approximately 0.5-in (0.6-cm) displacement of the entire second block course toward B-drift for X-4 stopping (Figure 70) and no significant displacement of this second block course for the X-5 stopping. A hairline horizontal crack was also observed near the roof above the top block course for the X-4 stopping.

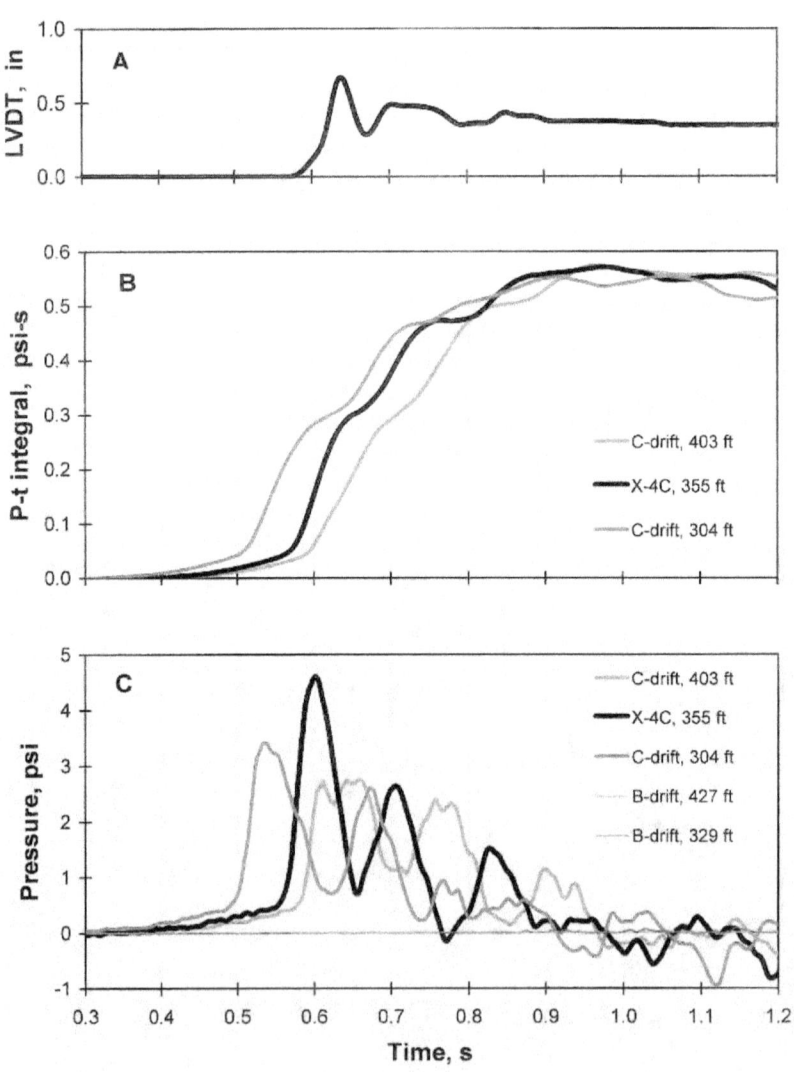

Figure 69.—(A) LVDT displacement, (B) pressure-time integrals, and (C) pressures versus time at the dry-stacked solid-concrete-block stopping in X-4 during LLEM test #459.

Figure 70.—Displacement of the second block course at the X-4 dry-stacked solid-concrete-block stopping after LLEM test #459, viewed from C-drift.

For all of the dry-stacked hollow-core and solid-concrete-block stoppings discussed in this report, a small amount of mortar was used under the first course of block to assist in the leveling of this first course in the sloped entry. Since the first block course was mortared to the concrete mine floor, it was expected that any block movement due to the explosion pressure would initially occur above the first course.

Test #460

During test #460 (Table A-11), the total explosion pressure was 6.7 psi (46 kPa) at the X-4 dry-stacked solid-concrete-block stopping. At the X-5 dry-stacked solid-concrete-block stopping, the total explosion pressure was 4.7 psi (33 kPa). The LVDTs showed movements of about 1.2 in (30 mm) for the X-4 stopping and about 0.8 in (20 mm) for the X-5 stopping. These LVDT displacements were recorded ~0.02 sec after the occurrence of the peak total explosion pressure on each stopping. This time delay may be due to the inertia of the stopping. After the explosion test, both stoppings were essentially intact. However, pronounced cracking was evident on the X-4 stopping between the first and second block course, with a 0.5- to 1-in (1.3- to 2.5-cm) total displacement of the entire second block course toward B-drift. Additional horizontal cracking above the entire top block course was noted, and a new hairline crack extended nearly rib to rib across the centerline of the stopping. For the X-5 stopping, a pronounced crack was observed between the first and second courses, with a 0.25-in (0.6-cm) displacement of the entire second course toward B-drift.

Test #461

During test #461, the total explosion pressures at the stoppings were not higher than those during test #460. The total explosion pressure was 4.2 psi (29 kPa) at the X-4 dry-stacked solid-concrete-block stopping, while the total explosion pressure was 4.0 psi (28 kPa) at the X-5 dry-stacked solid-concrete-block stopping. The LVDTs showed movements of 0.6 in (15 mm) for the X-4 stopping and about 0.3 in (8 mm) for the X-7 stopping. No significant additional damage to the stoppings was observed following this explosion.

Test #462

During test #462, the total explosion pressure was 7.6 psi (52 kPa) at the X-4 dry-stacked solid-concrete-block stopping (graph C of Figure 71). This stopping was essentially destroyed. Only a few blocks remained in place along the outby rib line and on the floor near the outby rib line (Figure 72). Approximately 2 ft (0.6 m) of the stopping block remained in place along the inby rib line and the floor near the inby rib line (Figure 73). Many of these remaining blocks from the original stopping location showed evidence of shearing failure (Figures 72–73). A map of the debris field from the destroyed X-4 stopping is shown in Figure 74. B-drift is at the lower part of the figure, and A-drift is at the top of the figure. The original position of the stopping is shown by the double horizontal line in X-4 near the bottom of the figure. Most of the stopping blocks were scattered to and beyond the B-drift intersection, with a few blocks thrown to the far wall of A-drift.

Figure 75 shows the LVDT displacement, pressure-time integrals, and pressures versus time at the X-5 dry-stacked solid-concrete-block stopping during test #462. As is shown in graph C of Figure 75 for the X-5 dry-stacked solid-concrete-block stopping, the total explosion pressure was 6.7 psi (46 kPa). The stopping withstood the pressure pulse. The stopping exhibited a more pronounced horizontal crack between the first and second block course, with an approximately 0.5-in (1.3 cm) displacement of the entire second block course toward B-drift. A new hairline crack extending from the right center of the stopping to the outby floor corner was also evident. At the center of the X-5 stopping, the LVDT showed movement of slightly over 0.9 in (23 mm) during the explosion, but returned close to its original position after the explosion (graph A of Figure 75). This displacement data may have been affected by the pressure coming through the inby crosscut (X-4 when that stopping was destroyed) and subsequently moving the fishing line extended between the X-5 stopping and the sensor in B-drift.

A summary of the pressure data at the dry-stacked solid-concrete-block stopping positions for test #462 is given in Figure 76 and Table 3. The maximum omnidirectional wall pressures in C-drift at the various DG panel positions are shown in Figure 76 as a function of distance from the face. At the top of Figure 76 is a schematic showing the locations of the stoppings. The stoppings in X-4 and X-5 were at distances of 355 and 451 ft (108 and 138 m) from the face, respectively. The B-drift pressures are the maximum values up until 0.1 sec after peak pressure in C-drift. (The pressures in B-drift were somewhat higher at later times due to the failure of the stoppings.) The total explosion pressures and pressure-time integrals at both of the dry-stacked solid-concrete-block positions are listed in Table 3.

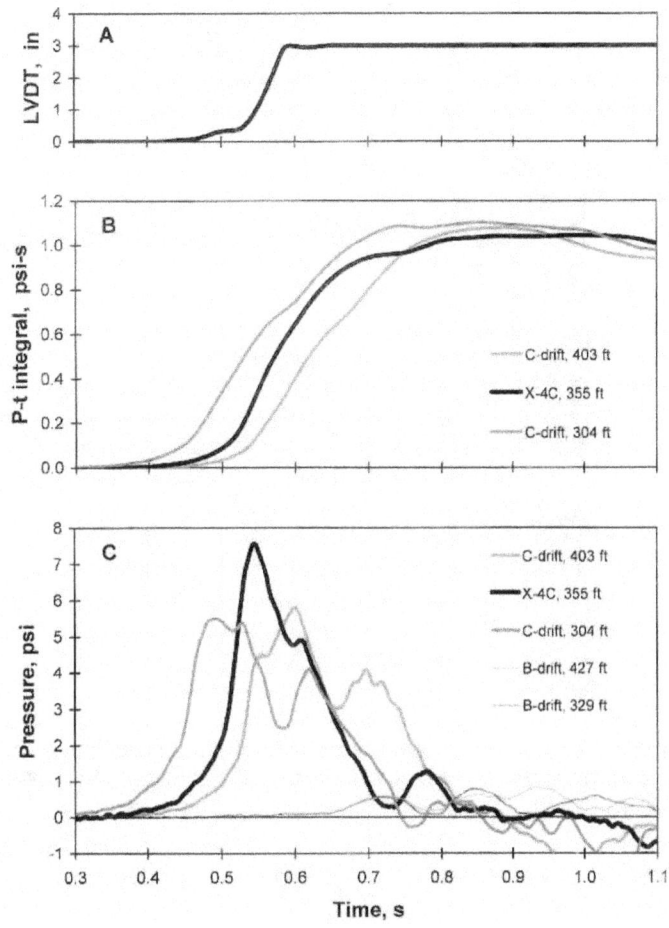

Figure 71.—(A) LVDT displacement, (B) pressure-time integrals, and (C) pressures versus time at the dry-stacked solid-concrete-block stopping in X-4 during LLEM test #462.

Figure 72.—Debris from the dry-stacked solid-concrete-block stopping and remaining blocks on the outby rib line in X-4 after LLEM test #462, viewed from C-drift.

Figure 73.—Debris from the dry-stacked solid-concrete-block stopping and remaining blocks on the inby rib line in X-4 after LLEM test #462, viewed from C-drift.

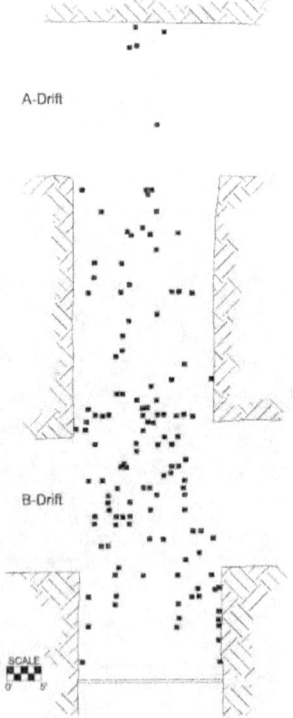

Figure 74.—Map of debris from the X-4 dry-stacked solid-concrete-block stopping after LLEM test #462. Black squares represent whole or partial blocks (smaller block pieces are not shown).

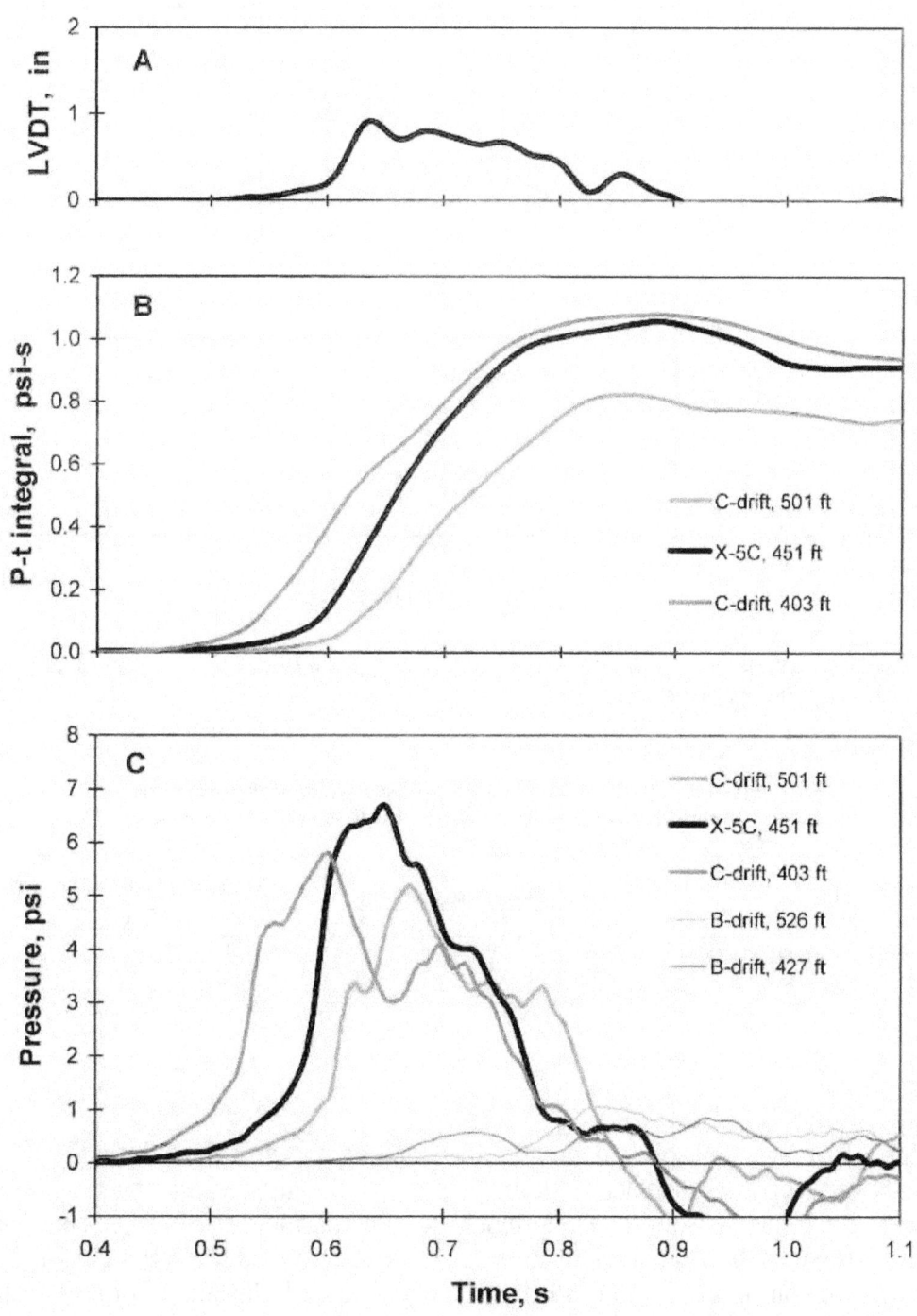

Figure 75.—(A) LVDT displacement, (B) pressure-time integrals, and (C) pressures versus time at the dry-stacked solid-concrete-block stopping in X-5 during LLEM test #462.

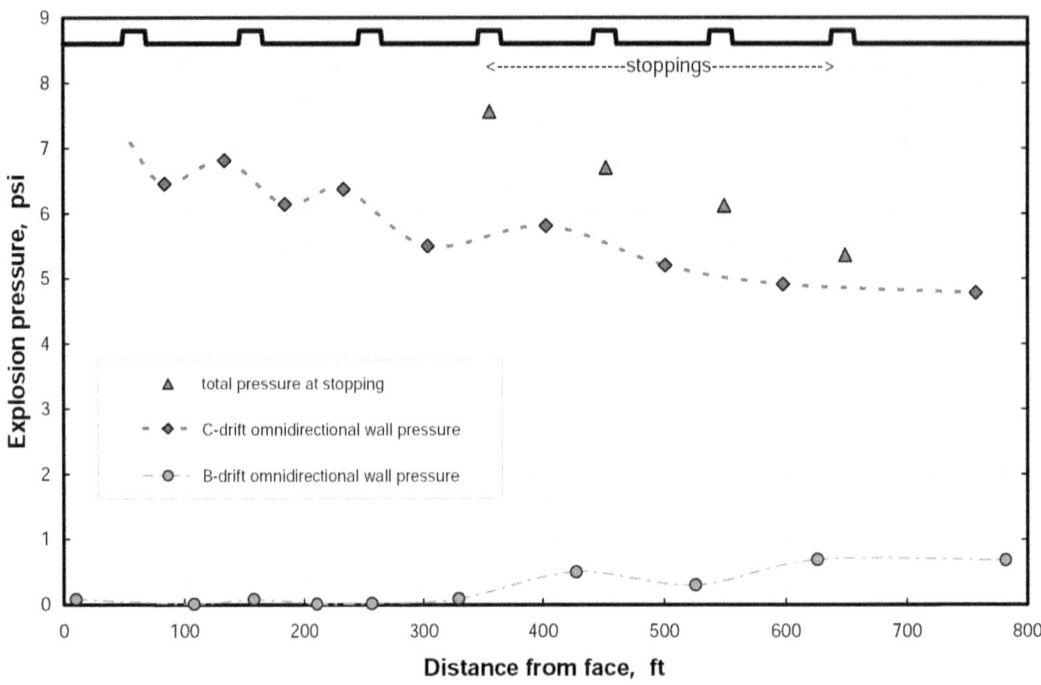

Figure 76.—Pressures at walls (ribs) and at the dry-stacked solid-concrete-block stopping locations during LLEM test #462.

Table 3.—Peak total explosion pressure data at dry-stacked solid-concrete-block stoppings for test #462 in C-drift

Location	Distance		Peak total explosion pressure		Pressure-time integral	
	ft	m	psi	kPa	psi-s	kPa-s
Crosscut 4	355	108	7.6	52	1.04	7.2
Crosscut 5	451	138	6.7	46	1.06	7.3

Test #463

The X-5 dry-stacked solid-concrete-block stopping was destroyed during test #463. Figure 77 shows the LVDT displacement, the pressure-time integrals, and the pressures versus time at the X-5 dry-stacked solid-concrete-block stopping during test #463. This test generated much higher pressures than the previous tests since it was designed to exert a 20-psi (138-kPa) explosion pressure pulse on a seal in X-1 that had been remotely installed from the surface through a borehole into the mine. During this test, the total explosion pressure was 16.6 psi (115 kPa) at the X-5 stopping (graph C of Figure 77). As shown in graph A of Figure 77, the LVDT also moved more than 3 in (8 cm) as the stopping was destroyed. Nearly all of the stopping blocks were scattered from the original stopping position except for a few blocks at the inby floor corner (Figure 78, which shows block shearing) and one block still intact at the outby floor corner. Figure 79 shows the debris map for the X-5 dry-stacked solid-concrete-block stopping after test #463. The stopping blocks were thrown into and beyond B-drift, with some blocks all the way into A-drift.

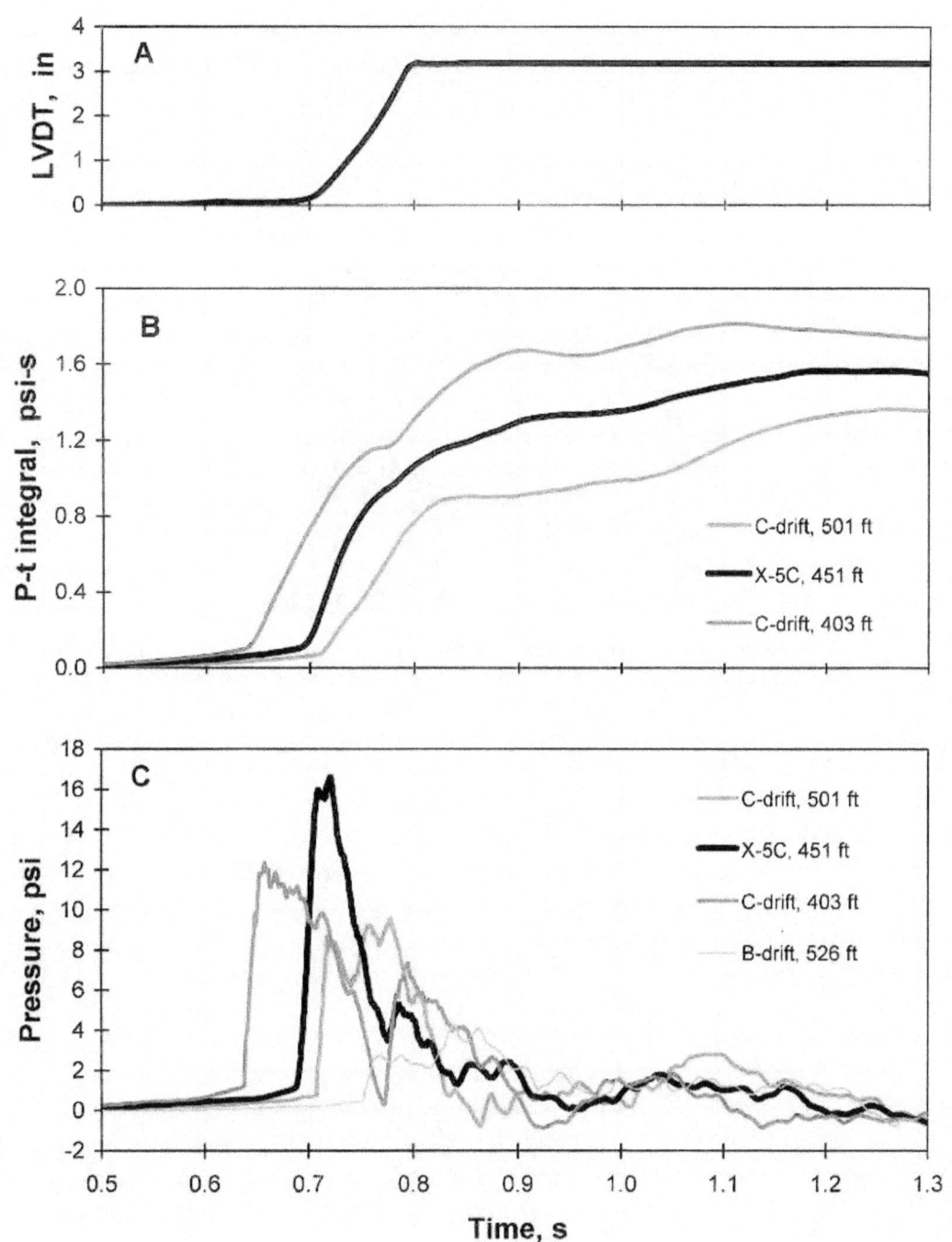

Figure 77.—(A) LVDT displacement, (B) pressure-time integrals, and (C) pressures versus time at the dry-stacked solid-concrete-block stopping in X-5 during LLEM test #463.

Figure 78.—Remaining blocks from the dry-stacked solid-concrete-block stopping at the inby rib in X-5 after LLEM test #463, viewed from B-drift.

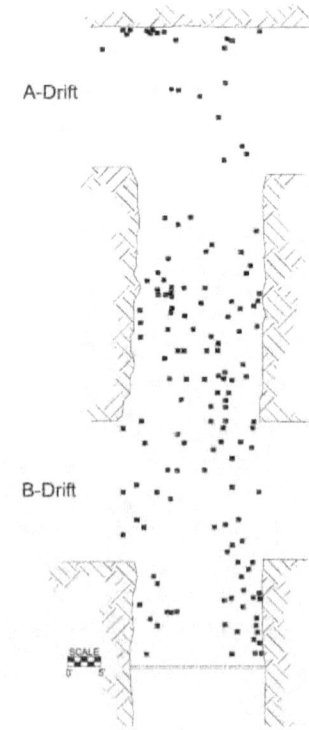

Figure 79.—Debris map for the X-5 dry-stacked solid-concrete-block stopping after LLEM test #463. Black squares represent whole or partial blocks (smaller block pieces are not shown).

Wet-Laid Solid-Concrete-Block Stoppings in Crosscuts

The two wet-laid solid-concrete-block stoppings in X-6 and X-7 between A- and B-drifts were evaluated during a series of 10 explosion tests (#510–#519) with total pressures ranging from 9.5 to 26.8 psi (66 to 185 kPa) at the X-6 stopping and 9.4 to 25.3 psi (65 to 174 kPa) at the X-7 stopping. A summary of three methane-only LLEM explosion tests (tests #510, #515, and #519) and one of the propagating coal dust explosion tests (test #512) is presented in Table 1. A complete listing of the A-drift omnidirectional wall pressures and stopping pressures for these three methane-only tests and the one test involving a coal dust explosion is also given in Tables A-20 through A-23 in the appendix.

Test #510

During the methane-only test #510, which consisted of a 40-ft (12-m) long gas ignition zone at the face of A-drift, there was little or no observable damage to the two wet-laid solid-concrete-block stoppings. The total explosion pressures were 14.5 psi (100 kPa) and 13.5 psi (93 kPa) at the stoppings in X-6 and X-7, respectively. The LVDTs showed movement toward B-drift of 0.3 in (7 mm) on the 8-in (20-cm) thick X-6 stopping and 0.5 in (13 mm) on the 6-in (15-cm) thick X-7 stopping. The pressures in B-drift behind the stoppings were very low (<0.6 psi (<4 kPa)) and were a result of the pressure traveling inby B-drift from the open end well past the time of the initial outward traveling pressure pulse in A-drift. The total explosion pressures and the pressure-time integrals at both of the wet-laid solid-concrete-block positions are listed in Table 4.

Table 4.—Peak total explosion pressure data at wet-laid solid-concrete-block stoppings for test #510 in A-drift

Location	Distance		Peak total explosion pressure		Pressure-time integral	
	ft	m	psi	kPa	psi-s	kPa-s
Crosscut 6	600	183	14.5	100	3.0	21
Crosscut 7	699	213	13.5	93	3.0	21

Tests #511–#514

Tests #511–#514 were mainly conducted to determine the rock dust inerting requirements of various sized coal dusts. The purpose of these tests was to determine whether the flame from a 40-ft (12.2-m) long gas ignition zone at the closed end of A-drift would continue to propagate through a 300-ft (91.4-m) long zone of a coal and rock dust mixture. The results from these dust tests will be published in a future report. During these four methane and coal dust explosion tests, there was no observable damage to the stoppings. The total explosion pressures ranged from 9.5 to 13.6 psi (66 to 94 kPa) at the X-6 stopping and 9.4 to 10.8 psi (65 to 75 kPa) at the X-7 stopping.

Test #515

Figures 80–81 show the pressure data at the wet-laid solid-concrete-block stopping positions for test #515, which used a 50-ft (15.2-m) long gas ignition zone and no coal dust. The X-6 and X-7 stoppings were both subjected to a total explosion pressure of 14.2 psi (98 kPa). The

LVDTs showed movement of 0.3 in (7 mm) on the 8-in (20-cm) thick X-6 stopping and 0.7 in (18 mm) on the 6-in (15-cm) thick X-7 stopping. The total explosion pressures and the pressure-time integrals at both of the wet-laid solid-concrete-block positions are listed in Table 5.

Table 5.—Peak total explosion pressure data at wet-laid solid-concrete-block stoppings for test #515 in A-drift

Location	Distance		Peak total explosion pressure		Pressure-time integral	
	ft	m	psi	kPa	psi-s	kPa-s
Crosscut 6	600	183	14.2	98	3.9	27
Crosscut 7	699	213	14.2	98	3.9	27

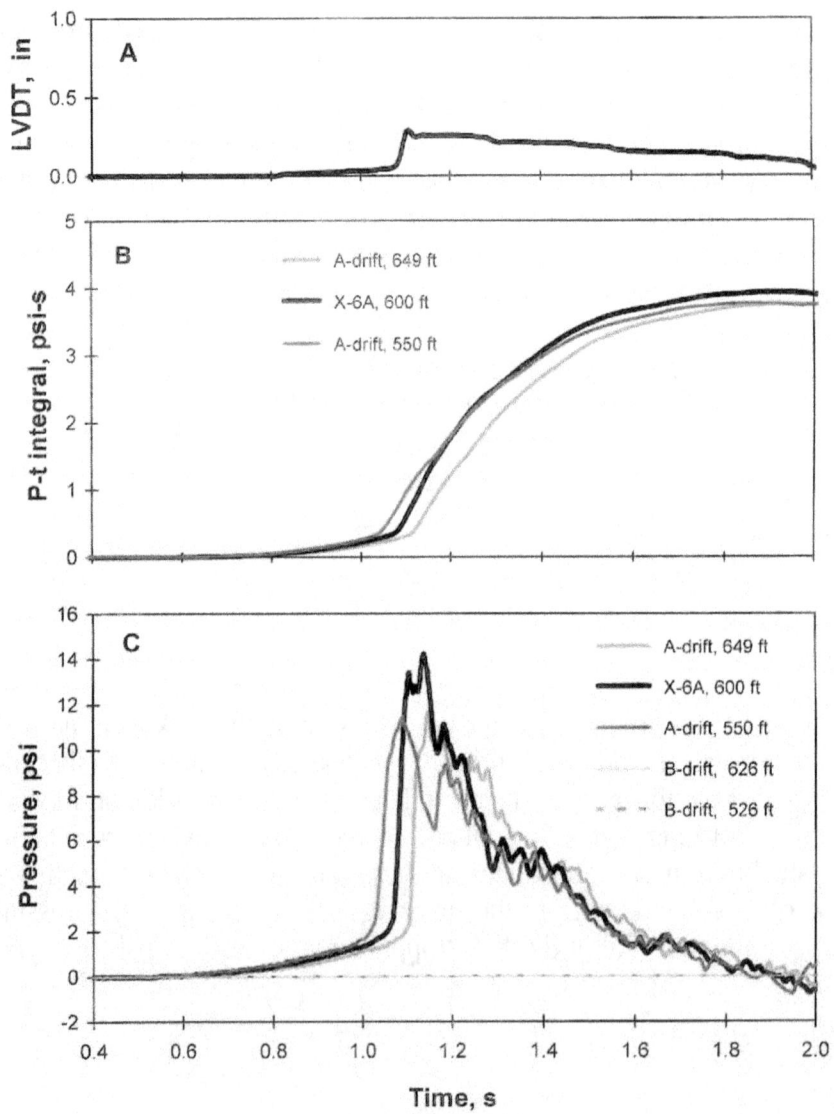

Figure 80.—(A) LVDT displacement, (B) pressure-time integrals, and (C) pressures versus time at the 8-in (20-cm) thick, wet-laid solid block stopping in X-6 during LLEM test #515.

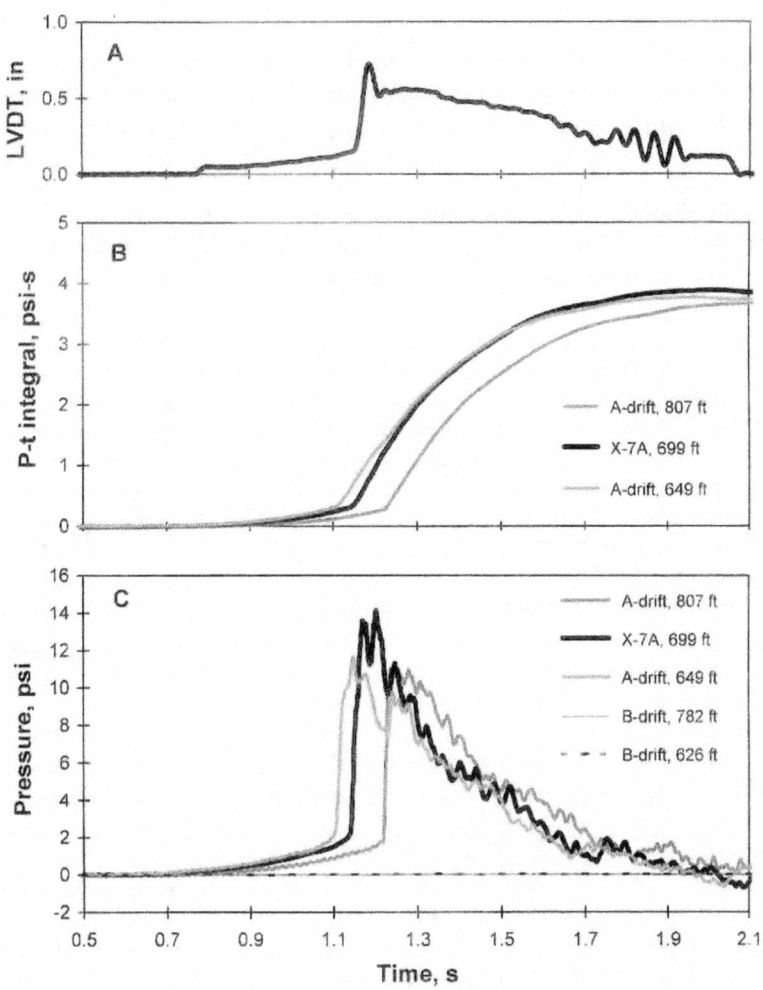

Figure 81.—(A) LVDT displacement, (B) pressure-time integrals, and (C) pressures versus time at the 6-in-thick, wet-laid solid block stopping in X-7 during LLEM test #515.

Tests #516–#518

Tests #516–#518 again involved coal and rock dust mixtures. During these tests, the explosion pressures ranged from 11.0 to 12.5 psi (76 to 86 kPa) at the X-6 stopping and 10.0 to 10.6 psi (69 to 73 kPa) at the X-7 stopping. The LVDTs showed movement of ~0.3 in (~8 mm) on the 8-in (20-cm) thick X-6 stopping and ~0.6 in (~18 mm) on the 6-in (15-cm) thick X-7 stopping during these tests. Small horizontal center cracks were first observed on the 6-in (15-cm) thick X-7 stopping after test #516. The 8-in (20-cm) thick X-6 stopping did not show any outward signs of damage during these tests.

Test #519

For test #519, there was an 85-ft (26-m) long gas zone at the face of A-drift, with no added coal dust. A summary of the pressure data at the wet-laid solid-concrete-block stopping positions for test #519 is given in Figures 82–83. The X-6 and X-7 stoppings were subjected to a total explosion pressure of 26.8 psi (185 kPa) and 25.3 psi (174 kPa), respectively. The LVDT showed movement of 1.0 in (26 mm) on the 8-in (20-cm) thick X-6 stopping, and a small horizontal center crack was observed on the A-drift side of the stopping after the test. The 6-in (15-cm) thick, wet-laid solid-concrete-block stopping in X-7 was destroyed during the test. Figure 84 shows the debris from the 6-in (15-cm) thick stopping in X-7 after test #519. The debris extended beyond B-drift to the far wall of C-drift. The block showed evidence of shearing. The total explosion pressures and the pressure-time integrals at both of the wet-laid solid-concrete-block stopping positions are listed in Table 6.

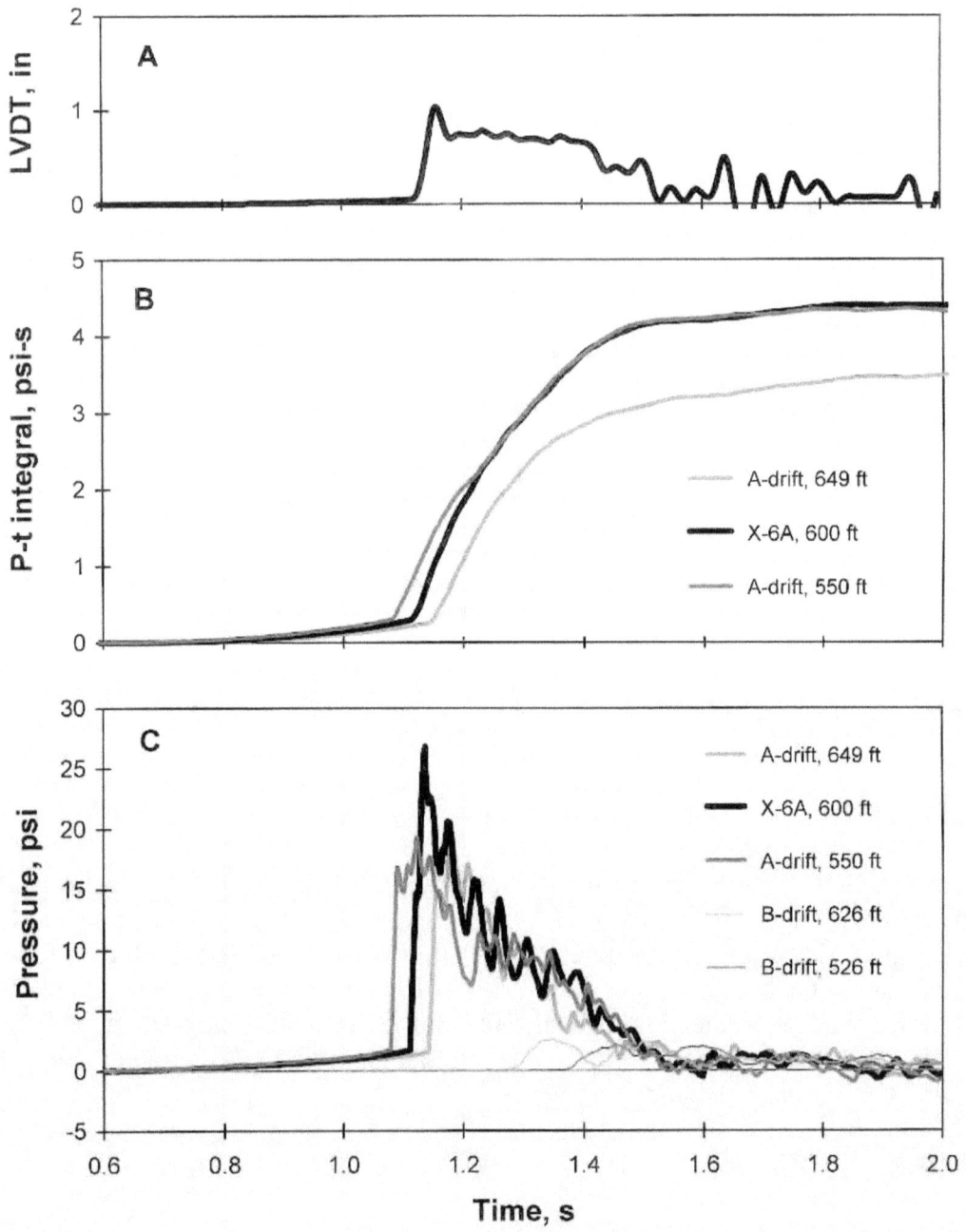

Figure 82.—(A) LVDT displacement, (B) pressure-time integrals, and (C) pressures versus time at the 8-in (20-cm) thick, wet-laid solid block stopping in X-6 during LLEM test #519.

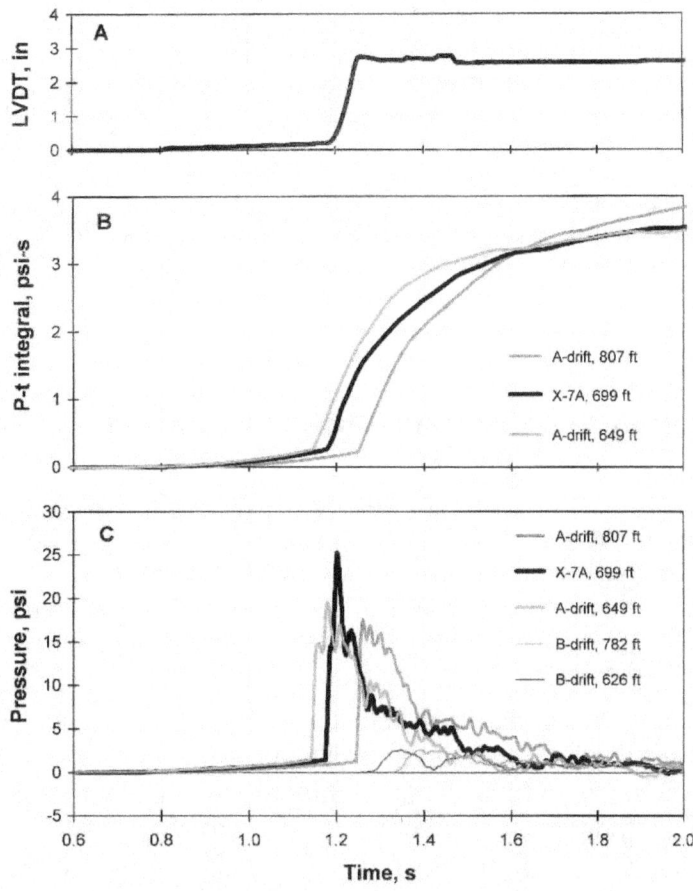

Figure 83.—(A) LVDT displacement, (B) pressure-time integrals, and (C) pressures versus time at the 6-in (15-cm) thick, wet-laid solid block stopping in X-7 during LLEM test #519.

Figure 84.—Debris from 6-in (15-cm) thick, wet-laid solid block stopping in X-7 after LLEM test #519, viewed from A-drift.

Table 6.—Peak total explosion pressure data at wet-laid solid-concrete-block stoppings for test #519 in A-drift

Location	Distance		Peak total explosion pressure		Pressure-time integral	
	ft	m	psi	kPa	psi-s	kPa-s
Crosscut 6	600	183	26.8	185	4.2	29
Crosscut 7	699	213	25.3	174	3.2	22

NOTE.—These stoppings were subjected to a total of 10 explosion tests, with pressures ranging from 9.5 to 26.8 psi at the X-6 stopping and 9.4 to 25.3 psi at the X-7 stopping.

Dry-Stacked Hollow-Core Concrete Block Stoppings in Entry and Crosscuts

MSHA investigations following methane and/or coal dust explosions in underground coal mines have documented cases where stoppings close to the ignition location have withstood the explosion pressure pulse, but stoppings farther away were destroyed.

To try to simulate this scenario, four and then later five dry-stacked hollow-core concrete block stoppings were constructed in the LLEM. The stoppings were located between A- and B-drifts and B- and C-drifts in X-3 and X-4. These stoppings were constructed in the same manner as previously described in the "Concrete Block Stoppings" section, except that sealant was only applied to one side of the stopping. Since many mines commonly coat stoppings on only one side, this series also provided an opportunity to evaluate the performance of these stoppings in comparison with those coated on both sides. The stoppings between A- and B-drifts were installed approximately 5 ft from A-drift (as measured from the closest A-drift rib) or about 35 ft from the closest B-drift rib line. The stoppings between B- and C-drifts were installed approximately 10 ft from the C-drift entry (as measured from the closest C-drift rib) or about 30 ft from the closest B-drift rib line. Later, a fifth stopping was installed across B-drift at 446 ft (136 m) from the face. A 2.6-ft (0.8-m) high by 3.8-ft (1.1-m) wide opening in the center of this stopping simulated an open regulator (Figure 85). All of the stoppings were coated with sealant on the B-drift (explosion) side. All of the blocks from each of the stoppings were sequentially numbered on the nonexplosion side of the stopping starting from the top. The blocks of each course were also designated with a letter, i.e., the blocks in the top course were labeled A-1, A-2, A-3, etc., the blocks in the second course from the top were labeled B-14, B-15, etc., as shown in Figure 86. A wire loop was installed on each stopping (nonexplosion side) to establish an electrical circuit or breakwire. This breakwire was designed to separate and thereby break the electrical circuit when the stopping was damaged or destroyed. The time of this event would be recorded and matched to the corresponding pressure peak. There were no LVDTs on these stoppings.

Figure 85.—Completed dry-stacked hollow-core concrete block stopping with simulated open regulator across B-drift.

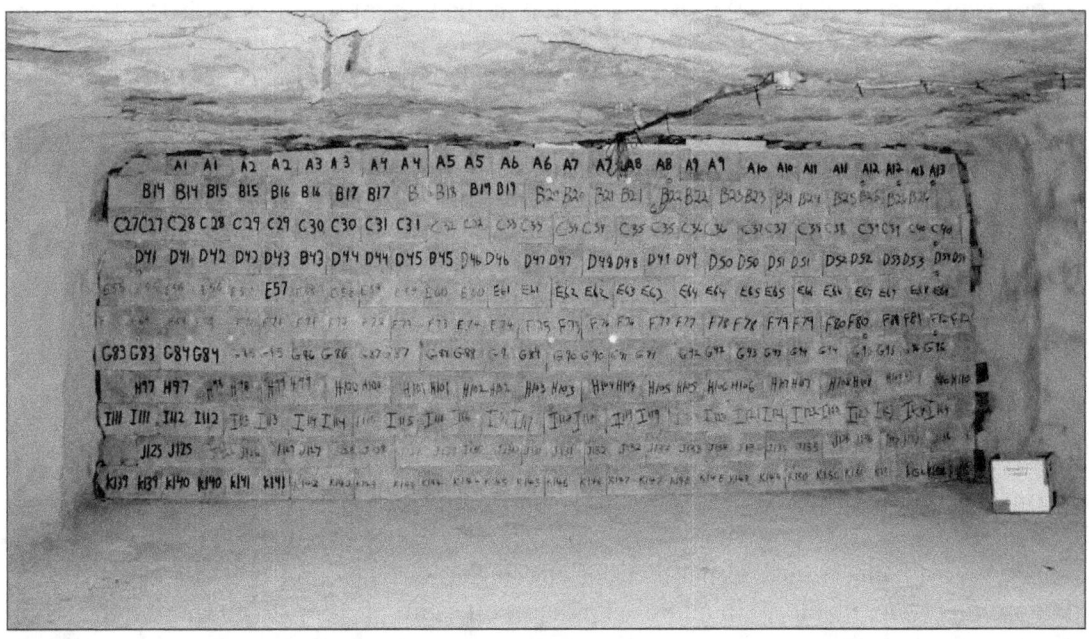

Figure 86.—Dry-stacked hollow-core concrete block stopping between B- and C-drifts in X-4 showing individually numbered block, viewed from C-drift.

Test #491

After installing the crosscut stoppings but before installing the stopping across B-drift, a short series of methane ignition experiments were conducted as part of another research program. It was not anticipated that the pressures generated from the ignition of a 38-ft (11.6-m) long, 10% methane-air concentration located between B-202 and B-240 would exceed 3 psi (21 kPa). This assumption was found to be incorrect. The ignition of this methane zone (LLEM test #491) generated total explosion pressures at the stopping locations ranging from 3.6 to 6.4 psi (25 to 44 kPa), which destroyed all four of the crosscut stoppings. The complete explosion pressure data for this test (LLEM test #491) are listed in Table A-15. These data are consistent with the results discussed in the "Dry-Stacked Hollow-Core Concrete Block Stoppings in Crosscuts" section, where similar dry-stacked hollow-core concrete block stoppings with coatings on both sides were destroyed at total explosion pressures between 3.4 psi (23 kPa) and 5.2 psi (36 kPa). The debris field of each stopping was mapped (Figure 87), and the information will assist in future accident investigations. Many of these remaining blocks from the original stopping location showed evidence of shearing failure.

Test #494

For test #494 (Table A-16), the four crosscut stoppings were rebuilt in addition to the stopping with an open regulator across B-drift at 446 ft (136 m) from the face. The methane ignition zone was reduced from 38 ft (11.6 m) long to 20 ft (6.1 m) long, located between B-240 and B-260 (between X-2 and X-3). The goal for this test was a lower-strength pressure pulse. Therefore, for test #494, a ~6% methane-air concentration was ignited by electric matches at the center of the 20-ft (6.1-m) long gas zone. The total explosion pressure was ~0.4 psi (~3 kPa) at the four crosscut stoppings and ~0.5 psi (~3.4 kPa) at the stopping across B-drift. Note that the listed pressure at the B-drift stopping was from a pressure transducer that was 18 ft (5.5 m) in front of the stopping or 428 ft (130 m) from the face. The pressure directly at this stopping may have been somewhat higher. All five of the dry-stacked hollow-core concrete block stoppings were still intact after the explosion test.

Test #495

During test #495, the gas concentration within the 20-ft zone was increased to ~7% methane-air. The total explosion pressure near the crosscut stoppings during this test (Table A-17) ranged from ~0.9 to ~1.1 psi (~6 to ~7.5 kPa). The total pressure was 1.2 psi (8.6 kPa) at the stopping across B-drift. All five of the stoppings were again intact after the explosion.

Test #496

During test #496, a ~9.5% methane-air concentration was ignited in the center of the 20-ft (6.1-m) gas zone, resulting in a total explosion pressure of 3.5 psi (24 kPa) at the stopping across B-drift at 446 ft from the face. All four of the crosscut stoppings were subjected to an explosion pressure pulse of ~3 psi (~21 kPa). A few hairline cracks in the sealant were observed on the stopping across B-drift and the stoppings in X-4. A complete listing of the B- and C-drift omnidirectional wall and stopping pressures for LLEM test #496 is given in Table A-18.

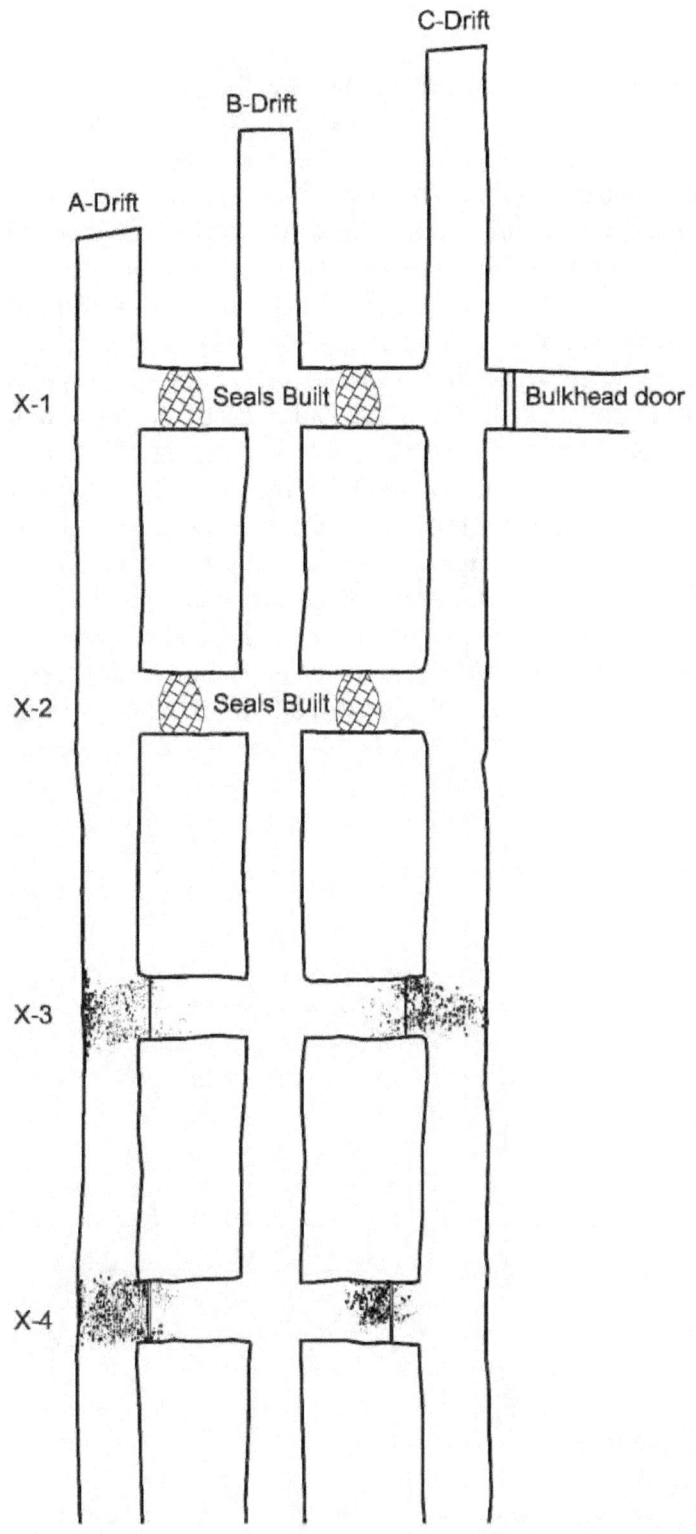

Figure 87.—Debris map after LLEM test #491 (overall view of entire area).

Test #497

During test #497, the length of the gas ignition zone was almost doubled to 38 ft (11.6 m) (located between B-202 and B-240). An ~8% methane-air concentration was ignited at the center of this 38-ft (11.6-m) long gas zone. The total explosion pressure on the B-446 stopping located across the drift was 5.2 psi (36 kPa) (see graph B of Figure 88 and Table A-19), which destroyed the stopping. Figure 89 shows the debris from the dry-stacked hollow-core concrete block stopping across B-drift; shearing of the blocks was also observed. The total explosion pressures near the stoppings in X-4 ranged from 3.6 to 3.8 psi (25 to 26 kPa), as shown in graph B of Figure 90. The X-4 stopping between B- and C-drifts was mostly destroyed, and the X-4 stopping between A- and B-drifts was totally destroyed, as shown in Figures 91 and 92, respectively. The total explosion pressure at the stoppings in X-3 was 4.3 psi (30 kPa), as shown in Figure 93. The stopping located between A- and B-drifts in X-3 was not damaged, and the stopping between B- and C-drifts in X-3 was only partly damaged (Figure 94). This seemed to be due to the failure of the coating since several individual blocks were knocked loose. A complete listing of the B- and C-drift omnidirectional wall and stopping pressures for this test is given in Table A-19. Figure 95 shows the debris map for all of the dry-stacked hollow-core concrete block stoppings after LLEM test #497. Three of the four seals that had been in X-1 and X-2 were removed prior to this test. A framed check curtain was installed in X-2 between B- and C-drifts, and it was knocked down by the explosion. These open crosscuts allowed additional venting of the pressures during this test.

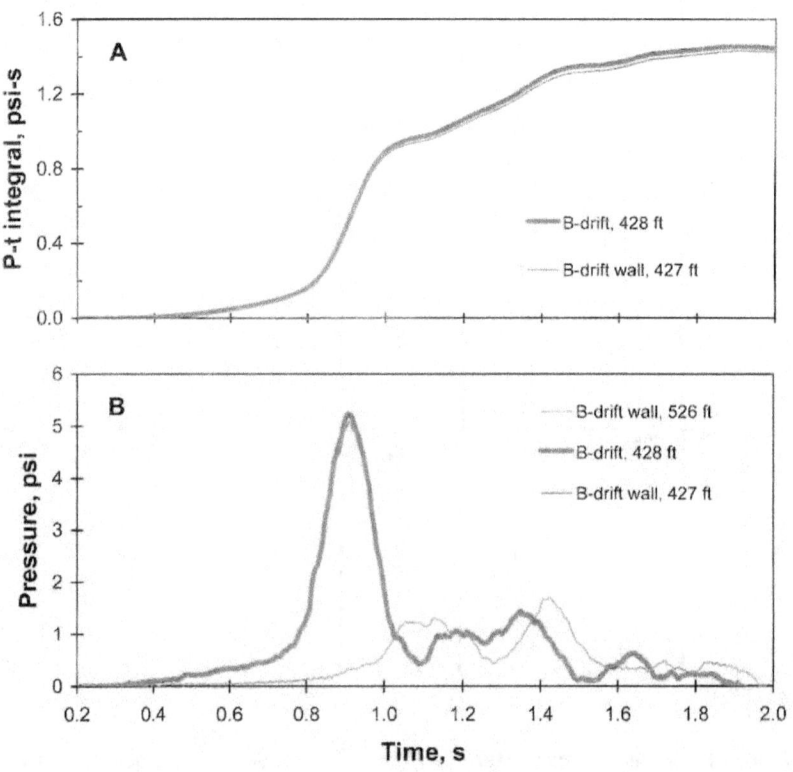

Figure 88.—(A) Pressure-time integrals and (B) pressures versus time at the dry-stacked hollow-core concrete block stopping across B-drift during LLEM test #497.

Figure 89.—Debris from the dry-stacked hollow-core concrete block stopping across B-drift after LLEM test #497.

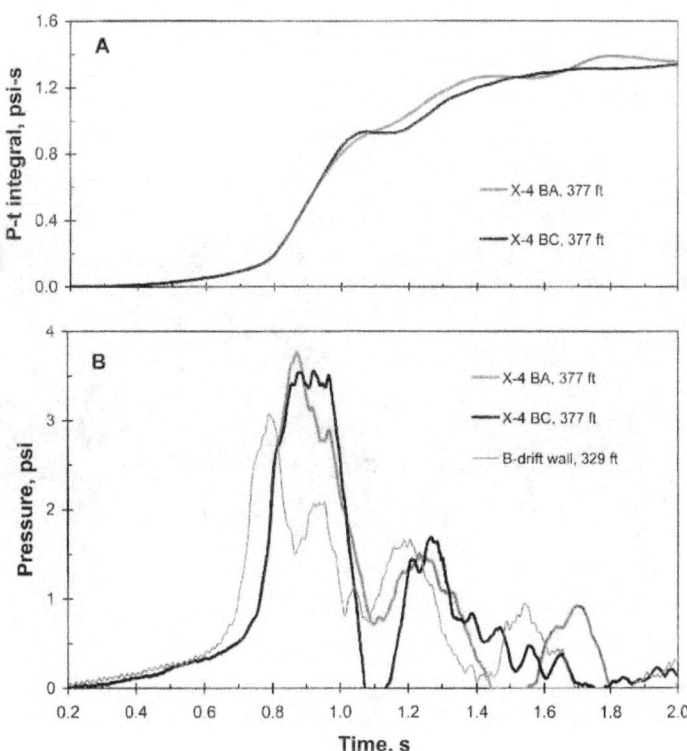

Figure 90.—(A) Pressure-time integrals and (B) pressures versus time at the X-4 dry-stacked hollow-core concrete block stoppings during LLEM test #497.

Figure 91.—Damage to the X-4 dry-stacked hollow-core concrete block stopping between B- and C-drifts after LLEM test #497.

Figure 92.—Debris from the X-4 dry-stacked hollow-core concrete block stopping between A- and B-drifts after LLEM test #497.

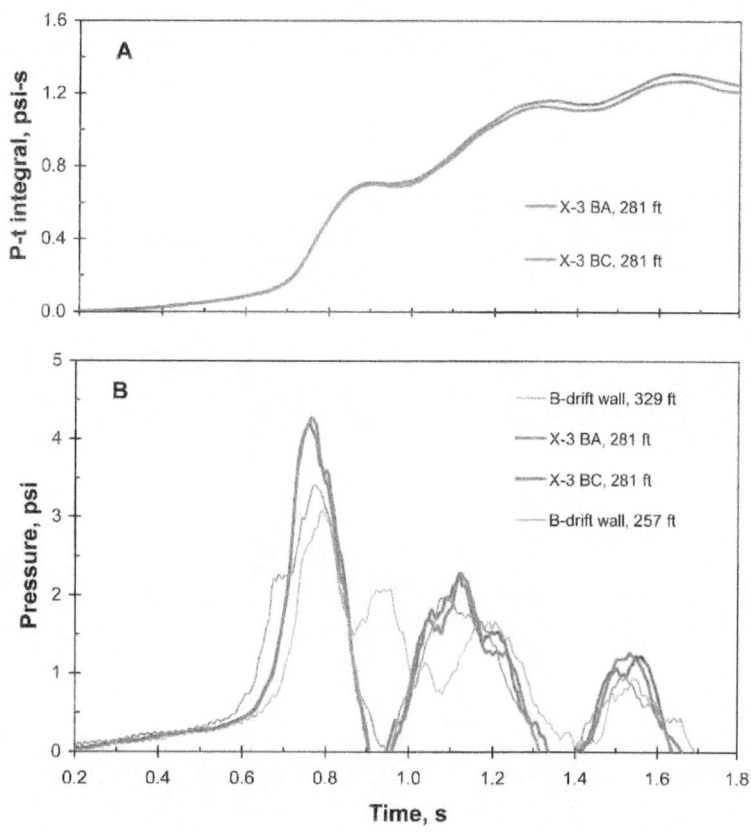

Figure 93.—(A) Pressure-time integrals and (B) pressures versus time at the X-3 dry-stacked hollow-core concrete block stoppings during LLEM test #497.

Figure 94.—Damage to the X-3 dry-stacked hollow-core concrete block stopping between B- and C-drifts after LLEM test #497.

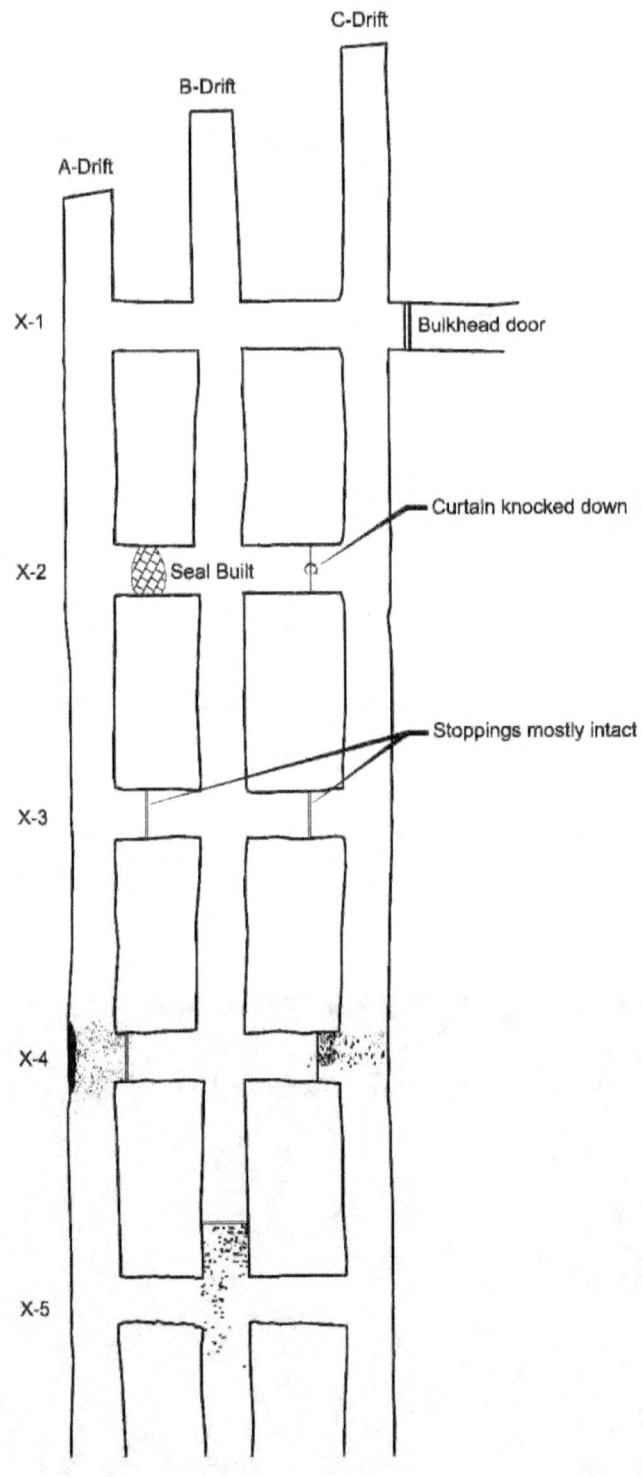

Figure 95.—Debris map after LLEM test #497 (overall view of entire area).

The results of test #497 demonstrated that a low-level pressure pulse (~3 psi (~21 kPa)) developed at the methane ignition zone (between X-2 and X-3) can move through an entry without damaging stoppings located in the nearby X-3. However, when that pressure pulse approaches an obstruction within the entry (a dry-stacked hollow-core concrete block stopping with an open regulator for this experiment), the pressure increases due to reflection at the B-drift stopping and thereby destroys the stopping. This reflected pressure increase is clearly evident on the B-257 wall pressure trace in Figure 93 where the initial outward traveling pressure pulse is ~2.3 psi (~16 kPa) and was subsequently increased to 3.4 psi (23 kPa) by the inward traveling reflected pressure pulse. The reflected pressure phenomenon is also shown on the B-drift 428-ft (130-m) pressure trace in Figure 96 where the initial outward traveling pressure pulse is ~2.3 psi (~16 kPa) at ~0.82 sec followed by a peak reflected pressure of 5.2 psi (36 kPa). The explosion pressure pulse also destroyed the inby X-4 stoppings at about the same time (Figure 96). These stoppings were recessed into X-4 between A- and B-drifts and between B- and C-drifts about the same distance that the stopping across B-drift was located outby of X-4. Only minor damage was evident on the X-3 stoppings. In this test, the stoppings in X-4 were destroyed at peak explosion pressures of 3.6–3.8 psi (25–26 kPa), while the stoppings in X-3 survived even though they were exposed to a peak pressure of 4.3 psi (30 kPa). However, the pressure-time integral was 0.93–0.94 psi-s (6.4–6.5 kPa-s) at the X-4 stoppings and only 0.70-0.71 psi-s (4.8–4.9 kPa-s) at the X-3 stoppings (Table A-19). This shows that the values of the pressure-time integral, as well as the peak total explosion pressure, are factors that affect whether a stopping will survive or be destroyed. In addition, the rate of pressure loading on a structure is also important in determining the response of the structure to an explosion.

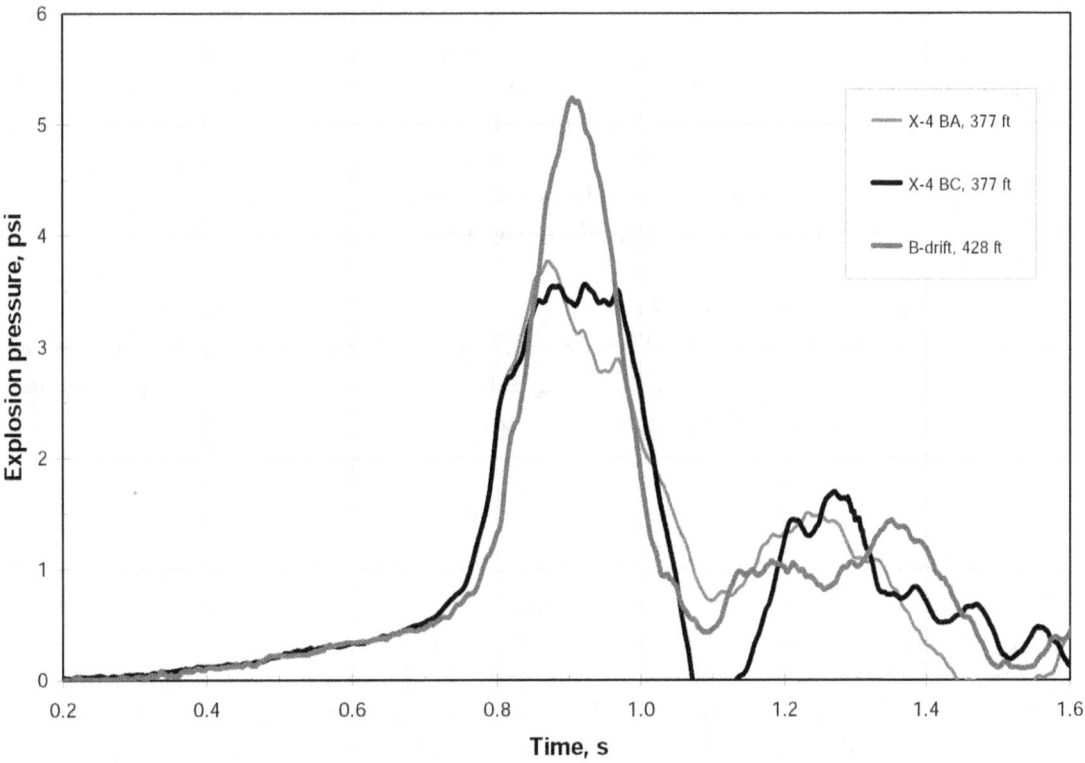

Figure 96.—Total explosion pressures versus time at the dry-stacked hollow-core concrete block stoppings in X-4 and in B-drift during LLEM test #497.

Australian Woven Cloth Stoppings in Crosscuts

Four full-scale explosion tests (LLEM tests #459–#462) were conducted in C-drift of the LLEM to evaluate the Australian-designed Flexi-Stop woven cloth stoppings in X-6 and X-7. These gas explosion tests were designed to provide an increasingly higher pressure pulse on the stopping designs during each subsequent test. A complete listing of the B- and C-drift omni-directional wall pressures and the stopping pressures for each test are given in the Tables A-10 through A-13 in the appendix. The Flexi-Stop stopping designs were located in X-6 at 547 ft (167 m) and X-7 at 647 ft (197 m) from the face of C-drift. Both stopping designs were eventually destroyed during the fourth explosion test.

A summary of these LLEM explosion tests is presented in Table 1. The first two tests had a 10-ft (3-m) deep by 12-ft (3.7-m) wide methane-air ignition zone contained in the face area by a clear plastic diaphragm. For the last two tests, the methane-air ignition zone was extended out to 27 ft (8.2 m) from the face.

Test #459

During test #459, there was very little observable damage to either of the two stopping designs. Figure 97 shows the pressure-time integrals and pressures versus time at the X-6 stopping during test #459; Figure 98 shows similar data at the X-7 stopping. The total explosion pressures, as measured ~4 ft (~1.2 m) directly in front of the stopping design, were 3.0 psi (21 kPa) at the X-6 stopping and 2.8 psi (19 kPa) at the X-7 stopping. For these evaluations, a pressure transducer was also mounted ~11 ft (~3.3 m) behind each stopping on the B-drift side. These transducers were centered within the crosscut and mounted so the sensor opening was perpendicular to any explosion gas flow or air displacement passing by the sensor location. In this orientation, the sensor would record the omnidirectional pressure. The pressure in B-drift behind each of the stoppings was initially zero but increased to about 0.1–0.2 psi (0.7–1.4 kPa). In Figures 97–98, the B-drift pressures recorded behind the X-6 and X-7 stoppings are designated as X-6B and X-7B, respectively. This slight pressure rise behind the stopping was most likely due to the compression of the air behind the stopping caused by the initial rapid billowing of the woven cloth toward B-drift during the explosion test. The recorded B- and C-drift pressure traces are shown in graph B of Figure 97 for the X-6 stopping and graph B of Figure 98 for the X-7 stopping. Graph A of Figures 97 and 98 shows the pressure-time integral for the X-6 and X-7 stoppings. The calculated pressure-time integral for the X-6 stopping was approximately 0.54 psi-s (4 kPa-s). For the X-7 stopping, the pressure-time integral was 0.55 psi-s (4 kPa-s).

Test #460

Prior to the second test against these Australian stopping designs (test #460), rib bolts were installed through the woven cloth and into each rib on each stopping (Figure 30). These rib bolts were designed as an additional anchoring technique to allow the stopping to withstand higher-level explosion pressures. Test #460 generated a total explosion pressure of 4.0 psi (27 kPa) for the X-6 design and 3.9 psi (27 kPa) for the X-7 stopping, with corresponding pressure-time integral values of 0.50 psi-s (3.4 kPa-s) for the X-6 stopping and 0.50 psi-s (3.5 kPa-s) for the X-7 stopping (Figures 99–100 and Table 7). Postexplosion observations revealed that the woven cloth was damaged (torn) along the rib lines at the locations of many of the rib bolt sites. It seemed that the square steel rib bolt plates provided a point loading contact area that ruptured the woven cloth starting at the plate corners. Figure 101 shows the damage to the woven cloth at the outby rib of X-6 after test #460 as viewed from C-drift. It can be readily seen in Figure 102 that the tears in the woven cloth on the inby rib of the X-7 stopping initiated at the bolt plate corners. A clear determination for the cause of the initial and subsequent pressure rises as measured from the transducer located behind each stopping was not readily discernable for this test or the subsequent tests. The pressure rises measured by these transducers could be a combination of two factors—displacement of the air column behind the stopping as resulting from the woven cloth billowing toward B-drift as the initial (and subsequent) explosion pressure pulse(s) impacted the C-drift side of the stopping and/or the pressure traveling through the tears in the woven cloth and/or damaged framework.

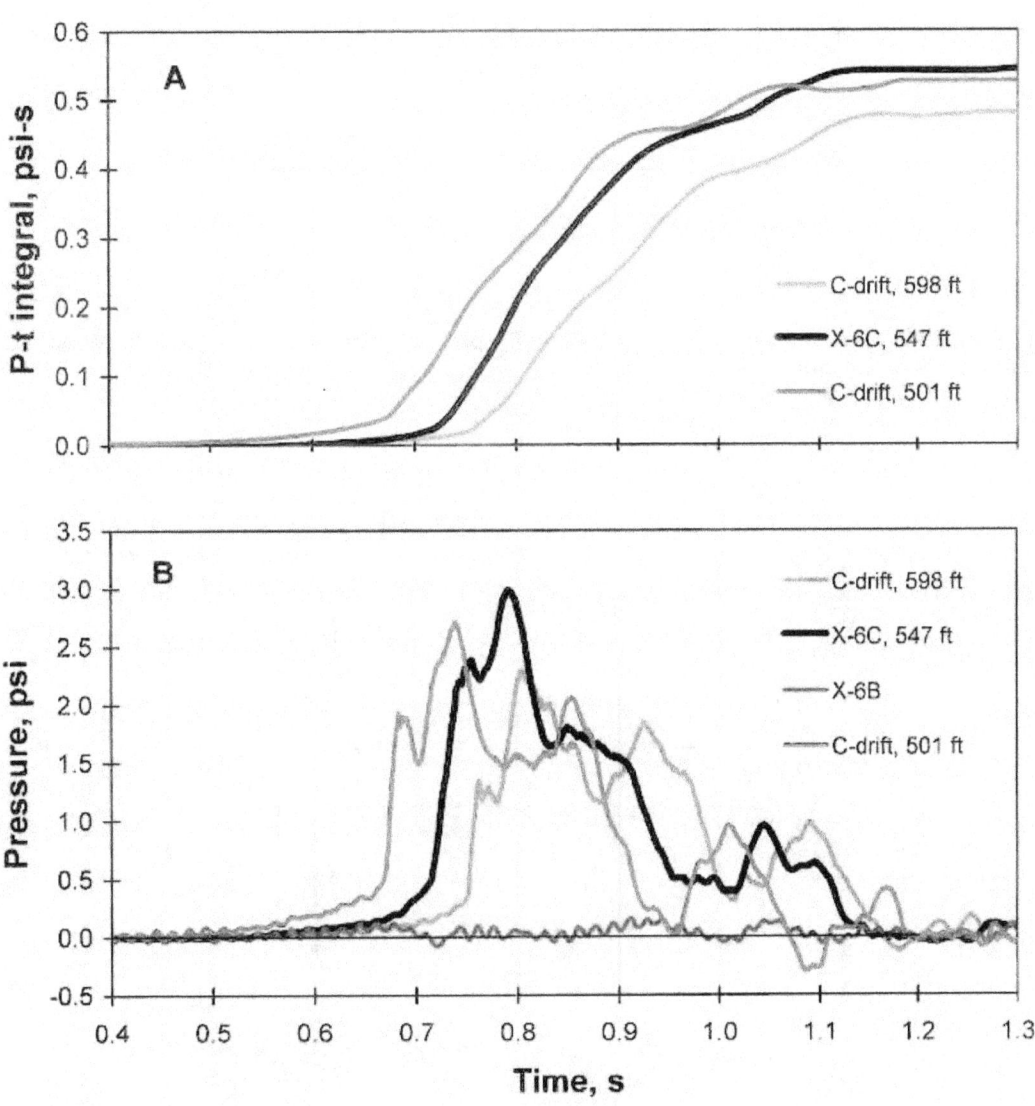

Figure 97.—(A) Pressure-time integrals and (B) pressures at the X-6 woven cloth stopping during LLEM test #459.

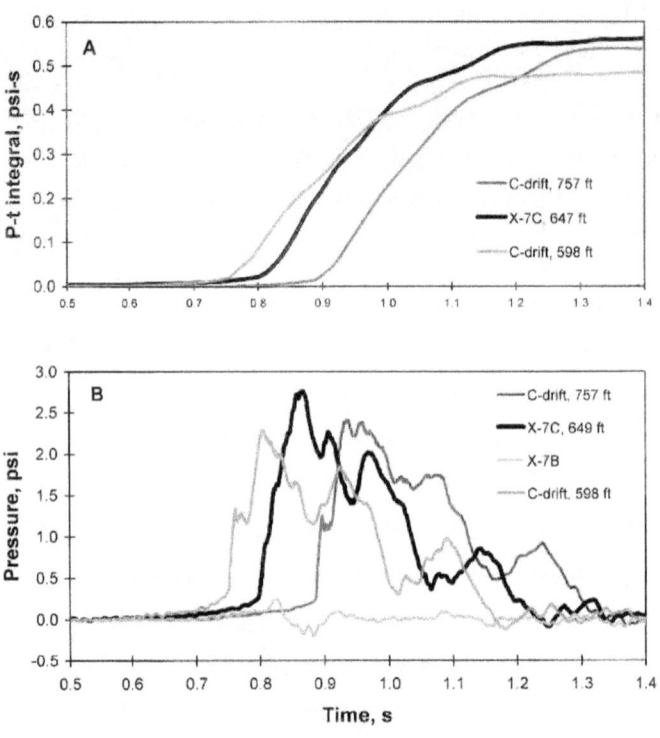

Figure 98.—(A) Pressure-time integrals and (B) pressures at the X-7 woven cloth stopping during LLEM test #459.

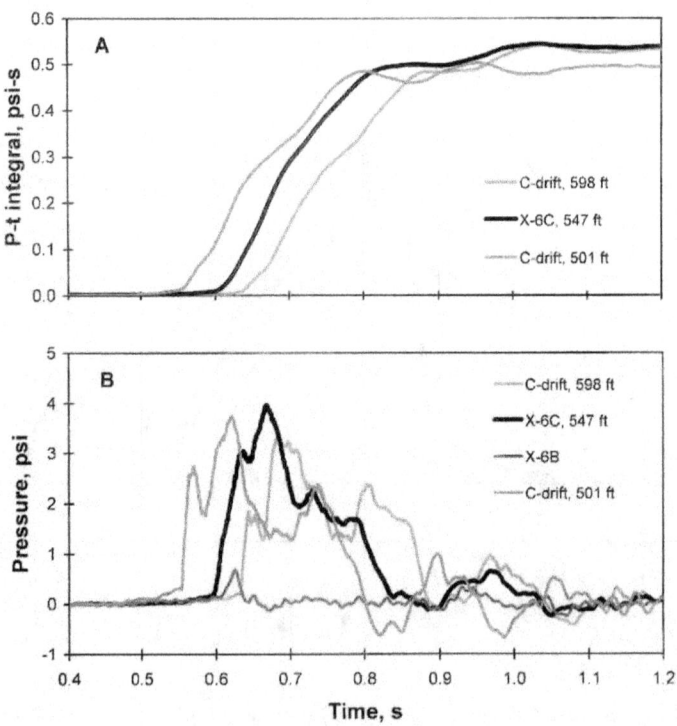

Figure 99.—(A) Pressure-time integrals and (B) pressures at the X-6 woven cloth stopping during LLEM test #460.

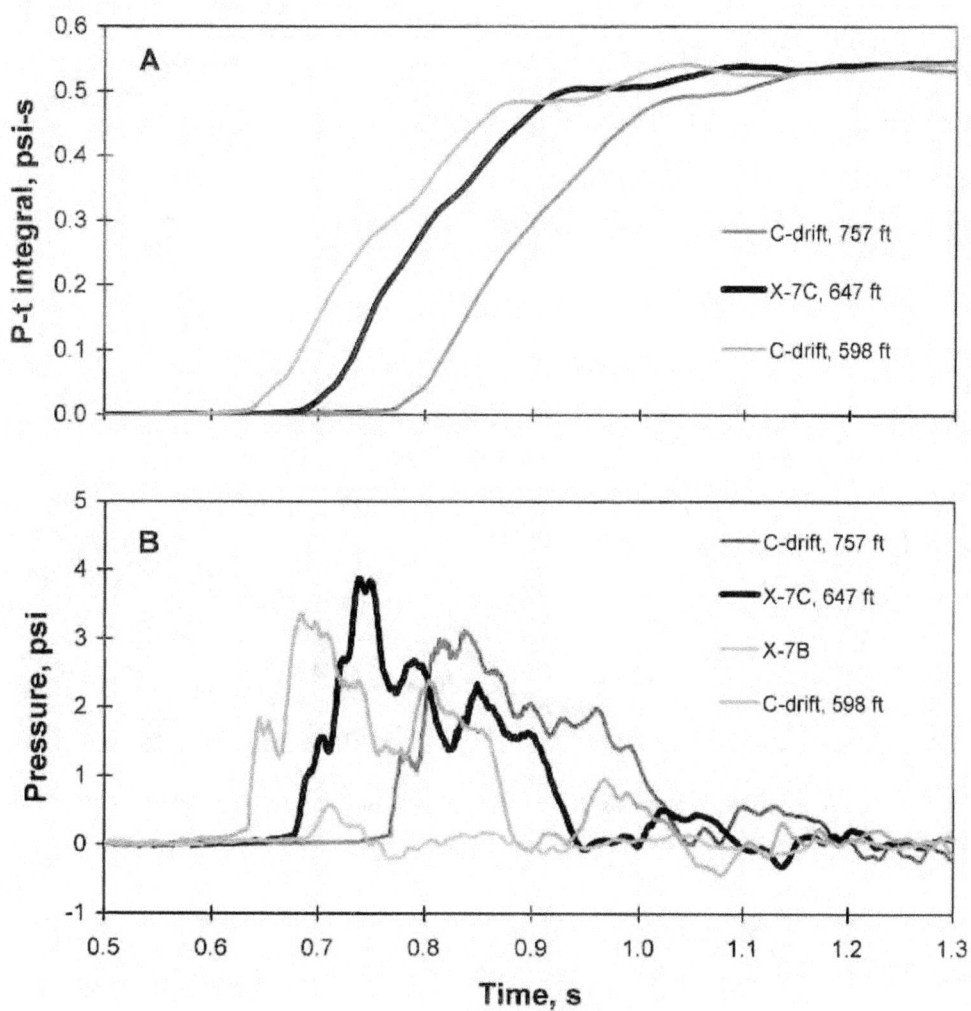

Figure 100.—(A) Pressure-time integrals and (B) pressures at the X-7 woven cloth stopping during LLEM test #460.

Table 7.—Peak total explosion pressure data at Australian woven cloth stoppings for test #460 in C-drift

Location	Distance		Peak total explosion pressure		Pressure-time integral	
	ft	m	psi	kPa	psi-s	kPa-s
Crosscut 6	547	167	4.0	27	0.50	3.4
Crosscut 7	647	197	3.9	27	0.50	3.5

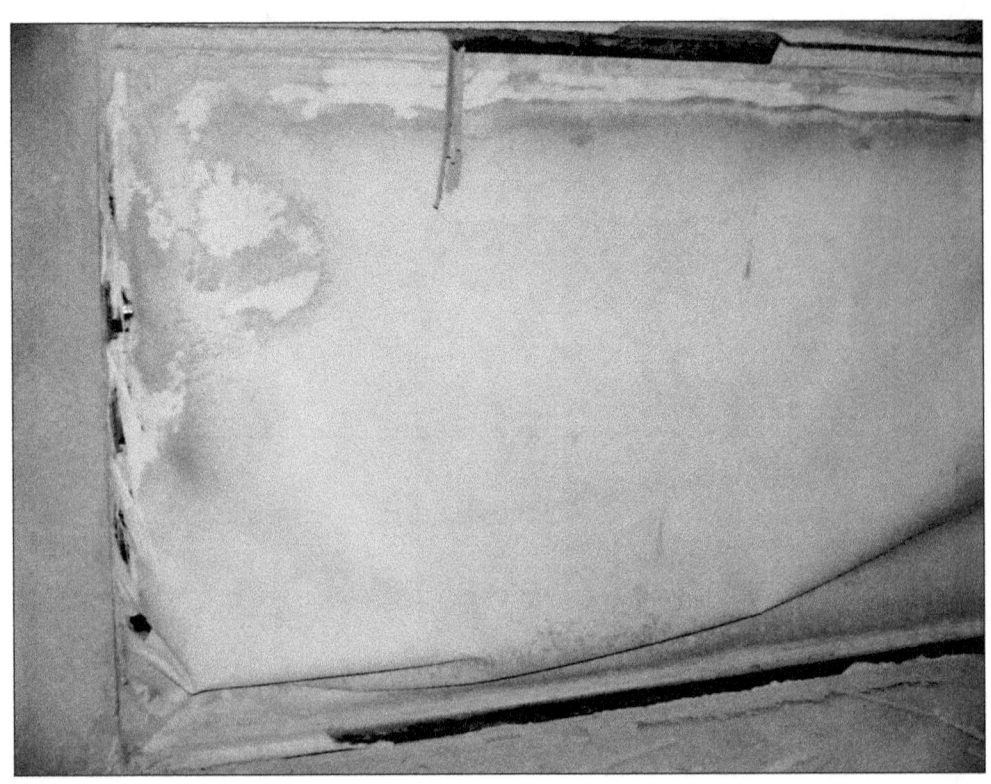

Figure 101.—Tears in the woven cloth on the outby rib of X-6 stopping after LLEM test #460, viewed from C-drift.

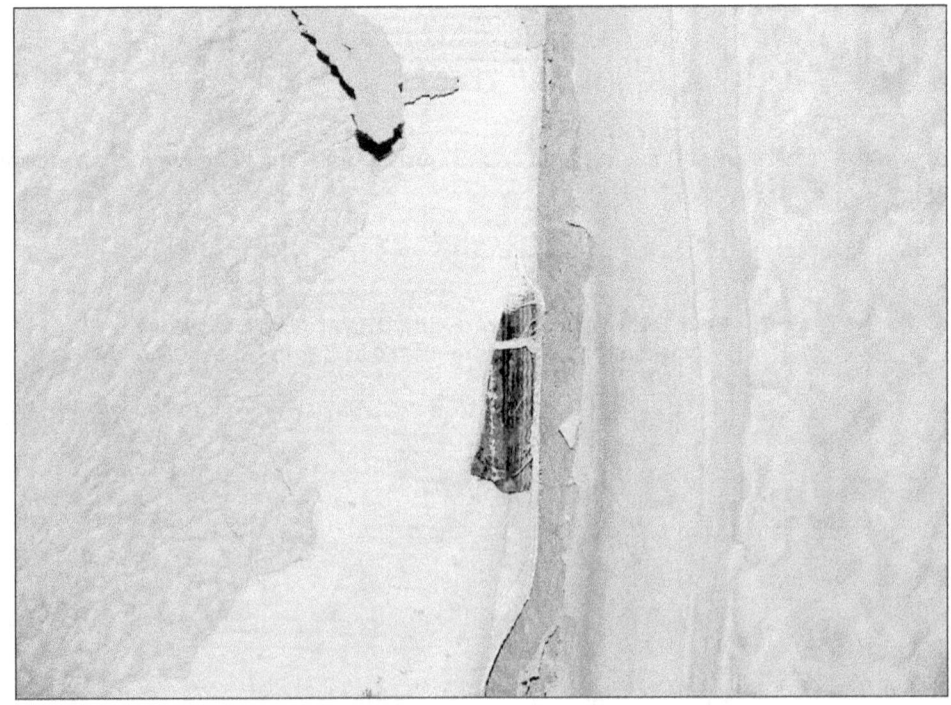

Figure 102.—Initial tear in the woven cloth at the corner steel roof bolt plate position on the inby rib of the X-7 stopping after LLEM test #460, viewed from C-drift.

Test #461

Test #461 was designed to provide a 5-psi (35-kPa) total explosion pressure at the stopping locations. However, the pressure levels were lower than expected. This may have been due to the improper mixing of the natural gas within the zone (the circulation fan was too small) and/or insufficient turbulence within the ignition zone to enhance the burning of the gas. The total explosion pressure was 3.7 psi (25.5 kPa) at the X-6 stopping and 3.6 psi (25 kPa) at the X-7 stopping. Postexplosion observations revealed very little indications of additional damage to the stoppings.

Test #462

For test #462, a double-point ignition source was used to initiate the methane, a larger circulation fan was used in the ignition zone, and the fan was left on during the ignition of the gas to provide increased turbulence. These changes were sufficient to provide the desired pressure loadings on the stoppings. The total explosion pressure was 6.1 psi (42 kPa) at the X-6 stopping and 5.4 psi (37 kPa) at the X-7 stopping, with corresponding pressure-time integral values of 0.71 psi-s (4.9 kPa-s) for the X-6 stopping and 0.85 psi-s (5.9 kPa-s) for the X-7 stopping (Table 8 and Figures 103–104). Postexplosion observations following test #462 revealed that the top steel box section and woven cloth for the X-6 stopping was on the floor (Figure 105). The four center bolts sheared at the roof and the steel rings on each end of the box section were sheared (roof bolts still anchored to the roof). The pipe/cloth assembly in the X-7 stopping pulled out from the slot in the top box section frame along a center 12-ft (3.5-m) portion of the frame. This resulted in an approximately 3-ft (1-m) opening between the mine roof and the now sagging woven cloth along most of the top section of the stopping (Figure 106). Both stoppings were rendered ineffective by the explosion.

Table 8.—Peak total explosion pressure data at Australian woven cloth stoppings for test #462 in C-drift

Location	Distance		Peak total explosion pressure		Pressure-time integral	
	ft	m	psi	kPa	psi-s	kPa-s
Crosscut 6	547	167	6.1	42	0.71	4.9
Crosscut 7	647	197	5.4	37	0.85	5.9

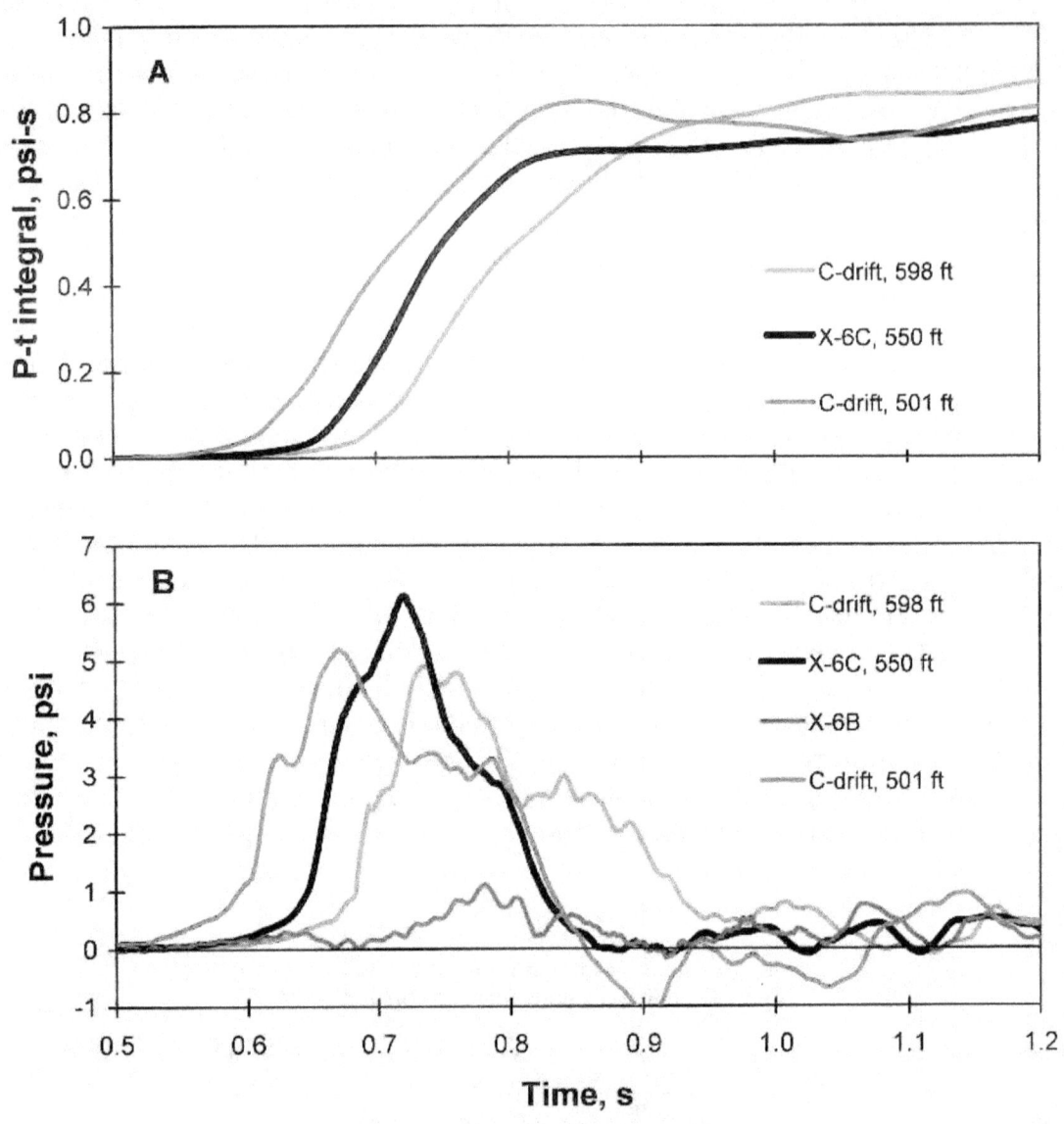

Figure 103.—(A) Pressure-time integrals and (B) pressures at the X-6 woven cloth stopping during LLEM test #462.

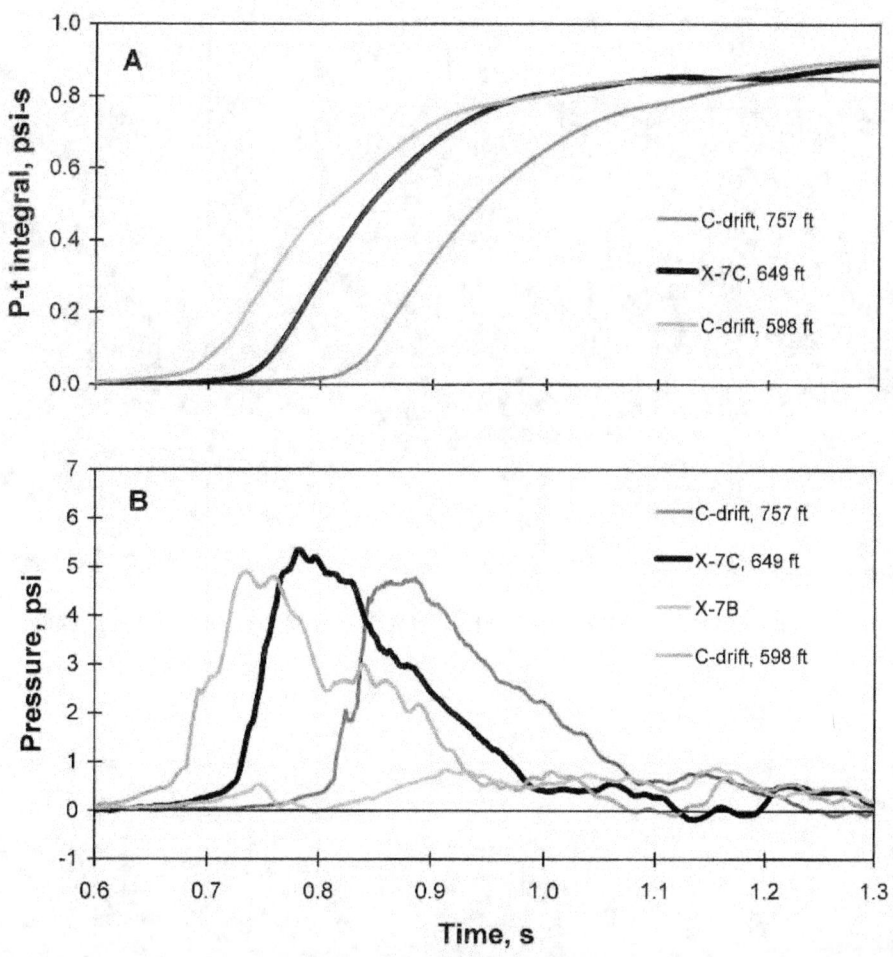

Figure 104.—(A) Pressure-time integrals and (B) pressures at the X-7 woven cloth stopping during LLEM test #462.

Figure 105.—Damage to the woven cloth stopping in X-6 after LLEM test #462.

Figure 106.—Damage to the woven cloth stopping in X-7 after LLEM test #462.

DISCUSSION AND CONCLUSIONS

NIOSH and MSHA conducted full-scale evaluation studies to determine the explosion pressures required to destroy typical coal mine ventilation stoppings in the LLEM. A summary of the total explosion pressure data for the various stoppings is shown in Table 9. A typical dry-stacked solid-concrete-block stopping, as installed in the LLEM for these tests, survived a total explosion pressure of ~6.7 psi (~46 kPa) and was destroyed at a total explosion pressure of ~7.6 psi (~52 kPa). In comparison, a typical dry-stacked hollow-core concrete block stopping survived a total explosion pressure of ~3.4–4.3 psi (~23–30 kPa), depending on the length of the pressure pulse and the value of the pressure-time integral. The dry-stacked hollow-core concrete block stopping was destroyed at a total explosion pressure of ~3.6–5.2 psi (~25–36 kPa), depending on the length of the pressure pulse and the value of the pressure-time integral. A typical steel panel stopping survived a total explosion pressure of 0.8 psi (5.5 kPa) and failed at a total explosion pressure of 1.3 psi (9 kPa). A 6-in (15-cm) thick wet-laid solid-concrete-block stopping survived a total explosion pressure of ~14 psi (~97 kPa) and was destroyed at a total explosion pressure of ~25 psi (~172 kPa). An 8-in (20-cm) thick wet-laid solid-concrete-block stopping survived a ~26-psi (~180-kPa) explosion pressure. The peak explosion pressures at which the concrete block and steel panel stoppings survived or failed was also dependent on the length of the pressure pulse, but there were insufficient data from these series of tests to quantify this. The value of the pressure-time integral, as well as the peak total explosion pressure and the pressure rise time, are factors affecting whether a stopping will survive or be destroyed. Note that for all of these stopping evaluations, the failure pressures are based on the construction conditions and explosion tests within the nonyielding limestone strata in the LLEM and the use of mortared floor joint to the concrete floor for the first course of blocks. Stopping strengths in mines may vary from these LLEM results when the stoppings are constructed in a coal seam and if subjected to roof convergence and/or floor heave or have significantly different boundary conditions.

Table 9.—Total explosion pressures necessary to destroy typical U.S. stoppings in the LLEM

Stopping type	Total explosion pressure at which stopping survived		Total explosion pressure at which stopping failed	
	psi	kPa	psi	kPa
Steel panel	0.8	5.5	1.3	9
6-in-thick dry-stacked hollow-core concrete block	~3.4–4.3	~23–30	~3.6–5.2	~25–36
6-in-thick dry-stacked solid concrete block	~6.7	~46	~7.6	~52
6-in-thick wet-laid solid concrete block	~14	~97	~25	~172
8-in-thick wet-laid solid concrete block	~26	~180	—	—

A NIOSH empirical correlation based on arching was recently developed by Barczak [2005] to determine the true transverse load capacity of dry-stacked concrete block walls. The load capacity would be the pressure that the stopping could withstand. In a mine, arching is achieved by the restraint of the block wall against the roof and floor, whereby compressive forces are developed within the wall. A systematic study of the design parameters that affect arching capability in dry-stacked concrete block walls was simulated in the NIOSH-PRL Mine Roof Simulator [Barczak and Batchler 2006]. The study included a theoretical assessment of arching and development of design formulations that can accurately define the transverse load capacity of dry-stacked concrete block wall constructions under various loading conditions. The correlation between three critical parameters (compressive strength (f_c), wall thickness (t), and wall height (H)) is shown (solid line) in Figure 107 as a function of the transverse load capacity. This correlation represents data from more than 70 tests including 8 different block materials, 4 nominal heights, 4 nominal thicknesses, and 10 concrete block compressive strengths. The term, $f_c(t/H)^2$, was derived from the moment equilibrium requirements of a half-wall loading [Barczak and Batchler 2006].

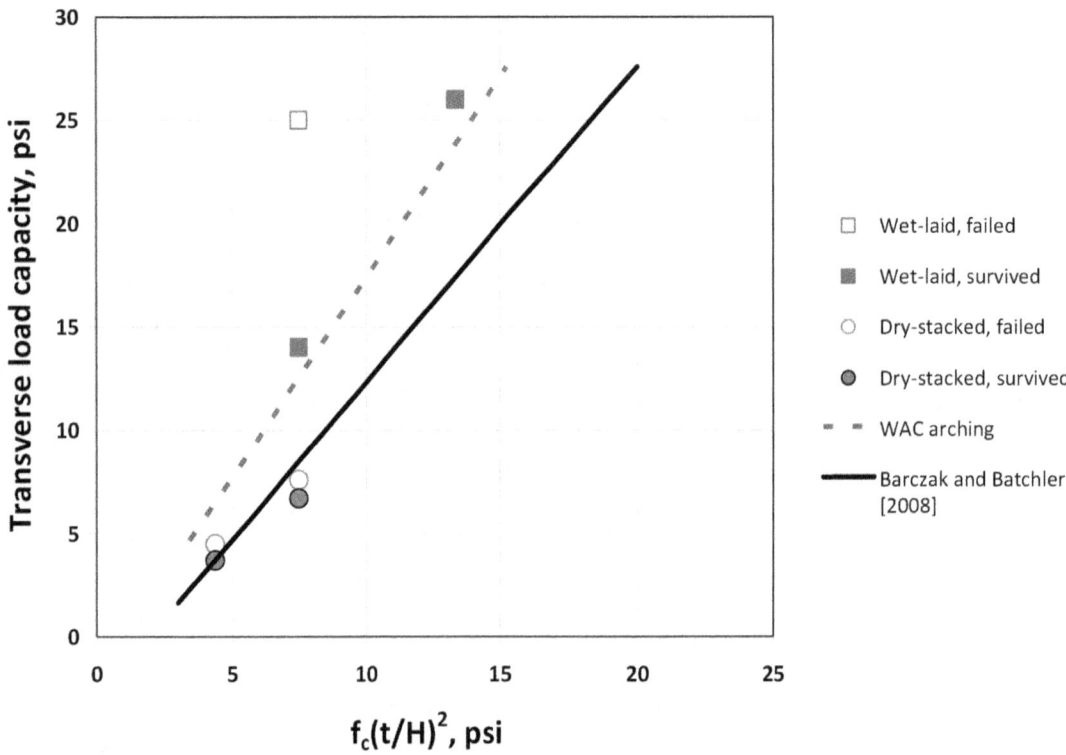

Figure 107.—Results of the LLEM block stopping evaluations compared to the critical design parameters (block compressive strength, wall height, and wall thickness) and the transverse load capacity predictions of the NIOSH empirical correlation by Barczak and Batchler [2008] (for dry-stacked block walls) and WAC (for wet-laid block walls) [Slawson 1995].

The strength of a stopping strongly depends on the boundary conditions around the perimeter of the stopping. Under rigid arch conditions, the lateral displacement of the wall is controlled by the stiffness and elastic response of the block wall. The transverse load capacity will decrease as the wall stiffness decreases since more lateral displacement will occur. The increase in lateral displacement reduces the force couple provided by the arching thrust, causing a decrease in the transverse load capacity of the stopping. If the abutments are not rigid, then the lateral displacement will increase further, resulting in a further reduction in the transverse load capacity of the stopping. The boundary condition in coal mines can vary significantly from one stopping location to another based on the perimeter strata conditions, the original stopping construction, and the subsequent degradation of the stopping and/or strata over time. The boundary conditions of the LLEM, with its limestone roof and ribs and concrete floor, are considered rigid abutment. The various data points for the LLEM stopping tests are shown in Figure 107. The LLEM explosion tests against the dry-stacked hollow-core concrete block stoppings and dry-stacked solid-concrete-block stoppings correlate well with the empirical correlations of Barczak and Batchler [2008]. The solid red circles in Figure 107 represent the dry-stacked concrete block stoppings that survived the LLEM explosion tests. The open circles represent the dry-stacked concrete block stoppings that failed the explosion tests.

Also shown in Figure 107 is the calculated static-elastic resistance for a wet-laid concrete block wall obtained from WAC [Slawson 1995] assuming one-way, roof-to-floor arching action and the same heights, thicknesses, and compressive strengths used in the NIOSH empirical correlations of Barczak and Batchler [2008]. The WAC-calculated static-elastic resistance for wet-laid concrete block walls (blue dashed line in Figure 107) shows a higher load capacity, as expected. The LLEM explosion tests against the wet-laid solid-concrete-block stoppings show reasonable agreement with the WAC-calculated predictions for wet-laid solid-concrete-block walls, although the LLEM data for the wet-laid stoppings are not as narrowly quantified as those for the dry-stacked stoppings. The blue solid squares represent the wet-laid solid-concrete-block stoppings that survived the LLEM explosion tests; the open square represents the one that failed the explosion test.

Explosion tests were also conducted in the LLEM that support observations from actual coal mine explosions where stoppings close to the ignition location can withstand a given explosion pressure impulse while stoppings farther away are destroyed, as discussed in the section entitled "Dry-Stacked Hollow-Core Concrete Block Stoppings in Entry and Crosscuts."

A program to evaluate the strength characteristics and air leakage resistance of an innovative Australian-designed woven cloth stopping showed that the stopping successfully withstood an explosion that generated a total explosion pressure of 4.0 psi (28 kPa) at the stopping location. The stopping failed when subjected to a subsequent explosion that generated a total explosion pressure of ~6.1 psi (~42 kPa) at the stopping location. Based on the results of this evaluation, this woven cloth stopping is now being used in underground coal mines in Australia.

The results from the LLEM stopping evaluations, coupled with the use of predictive wall strength models (which are based on the critical parameters including block strength, stopping height and thickness, and boundary conditions), can assist investigators in more accurately determining the range of explosion pressures that destroy or damage stoppings during actual coal mine explosions.

ACKNOWLEDGMENTS

The authors thank C. W. Moore and James Baca, Mining Engineers with the Ventilation Division, Pittsburgh Safety and Health Technology Center, MSHA, Triadelphia, WV, for their contributions to the mapping of the block stopping debris following several of the explosion evaluations.

We also thank the following NIOSH-PRL personnel without whose contributions this program could not have been accomplished: Kenneth W. Jackson, Electronics Technician, for sensor calibrations and installations, for operation of the data acquisition systems, and for his participation in the testing and data analyses; Frank A. Karnack, Donald D. Sellers, and

William A. Slivensky, Physical Science Technicians, for their participation in the installation of sensors and mounting equipment, the construction of the stoppings, testing, and cleanup; and John J. Glad (now retired), James D. Addis (now a Physical Science Technician with PRL), Timothy Glad, James Rabon, Bernie Lambie, and Brandin Lambie, Mechanical Technicians with Akima and Ki, for their installation of the concrete block and Australian stopping designs.

The authors acknowledge William Kennedy, President of Jack Kennedy Metal Products & Buildings, Inc., and Robert Cox, President of Ken-Air, Inc., for constructing the Kennedy steel panel stoppings.

We also acknowledge I. Verne S. Mutton, Senior Mining Engineer with Minova Australia, for his innovative, woven cloth coal mine stopping designs and his direct contributions to the installation of these designs in the LLEM.

All photographs in this report were taken by Kenneth L. Cashdollar, Eric S. Weiss, or Frank A. Karnack of NIOSH-PRL.

REFERENCES

61 Fed. Reg. 9764 [1996]. Mine Safety and Health Administration, 30 CFR part 75: safety standards for underground coal mine ventilation; final rule.

Barczak TM [2005]. Evaluation of the transverse load capacity of block stoppings for mine ventilation control [Dissertation]. Morgantown, WV: West Virginia University, Department of Mining Engineering.

Barczak TM, Batchler TJ [2006]. Development of new protocols to evaluate the transverse loading of mine ventilation stoppings. In: Mutmansky JM, Ramani RV, eds. Proceedings of the 11th U.S./North American Mine Ventilation Symposium (University Park, PA, June 5–7, 2006). London: Taylor & Francis Group, pp. 569–577.

Barczak TM, Batchler TJ [2008]. Comparison of the transverse load capacities of various block ventilation stoppings under arch loading conditions. In: Wallace KG Jr., ed. Proceedings of the 12th U.S./North American Mine Ventilation Symposium (Reno, NV, June 9–11, 2008). Reno, NV: University of Nevada, pp. 225–231.

CFR. *Code of Federal regulations.* Washington, DC: U.S. Government Printing Office, Office of the Federal Register.

Gates RA, Phillips RL, Urosek JE, Stephan CR, Stoltz RT, Swentosky DJ, Harris GW, O'Donnell JR Jr., Dresch RA [2007]. Report of investigation: fatal underground coal mine explosion, January 2, 2006. Sago mine, Wolf Run Mining Company, Tallmansville, Upshur County, West Virginia, ID No. 46-08791. Arlington, VA: U.S. Department of Labor, Mine Safety and Health Administration.

Greninger, NB, Weiss ES, Luzik SJ, Stephan CR [1991]. Evaluation of solid-block and cementitious foam seals. Pittsburgh, PA: U.S. Department of the Interior, Bureau of Mines, RI 9382. NTIS No. PB 92-152115.

Kawenski EM, Mitchell DW, Bercik GR, Frances A [1965]. Stoppings for ventilating coal mines. Pittsburgh, PA: U.S. Department of the Interior, Bureau of Mines, RI 6710.

Mattes RH, Bacho A, Wade LV [1983]. Lake Lynn Laboratory: construction, physical description, and capability. Pittsburgh, PA: U.S. Department of the Interior, Bureau of Mines, IC 8911. NTIS No. PB 83-197103.

McKinney R, Crocco W, Stricklin KG, Murray KA, Blankenship ST, Davidson RD, Urosek JE, Stephan CR, Beiter DA [2002]. Report of investigation: fatal underground coal mine explosions, September 23, 2001. No. 5 Mine, Jim Walter Resources, Inc., Brookwood, Tuscaloosa County, Alabama, ID No. 01-01322. Arlington, VA: U.S. Department of Labor, Mine Safety and Health Administration.

Mitchell DW [1971]. Explosion-proof bulkheads: present practices. Pittsburgh, PA: U.S. Department of the Interior, Bureau of Mines, RI 7581. NTIS No. PB 205 507.

Nagy J [1981]. The explosion hazard in mining. Pittsburgh, PA: U.S. Department of Labor, Mine Safety and Health Administration, IR 1119.

Slawson TR [1995]. Wall response to airblast loads: the Wall Analysis Code (WAC). Prepared for the U.S. Army Energy Research and Development Center, Vicksburg, MS.

Triebsch G, Sapko MJ [1990]. Lake Lynn Laboratory: a state-of-the-art mining research laboratory. In: Proceedings of the International Symposium on Unique Underground Structures. Vol. 2. Golden, CO: Colorado School of Mines, pp. 75-1 to 75-21.

Weiss ES, Greninger NB, Stephan CR, Lipscomb JR [1993]. Strength characteristics and air-leakage determinations for alternative mine seal designs. Pittsburgh, PA: U.S. Department of the Interior, Bureau of Mines, RI 9477. NTIS No. PB94111275.

Weiss ES, Slivensky WA, Schultz MJ, Stephan CR, Jackson KW [1996]. Evaluation of polymer construction material and water trap designs for underground coal mine seals. Pittsburgh, PA: U.S. Department of Energy, RI 9634. NTIS No. PB96-123392.

Weiss ES, Cashdollar KL, Mutton IVS, Kohli DR, Slivensky WA [1999]. Evaluation of reinforced cementitious seals. Pittsburgh, PA: U.S. Department of Health and Human Services, Public Health Service, Centers for Disease Control and Prevention, National Institute for Occupational Safety and Health, DHHS (NIOSH) Publication No. 99-136, RI 9647.

Weiss ES, Cashdollar KL, Sapko MJ [2002]. Evaluation of explosion-resistant seals, stoppings, and overcast for ventilation control in underground coal mining. Pittsburgh, PA: U.S. Department of Health and Human Services, Public Health Service, Centers for Disease Control and Prevention, National Institute for Occupational Safety and Health, DHHS (NIOSH) Publication No. 2003–104, RI 9659.

Weiss ES, Cashdollar KL, Harteis SP, Shemon GJ, Beiter DA, Urosek JE [2006]. Explosion evaluation of mine ventilation stoppings. In: Mutmansky JM, Ramani RV, eds. Proceedings of the 11th U.S./North American Mine Ventilation Symposium (University Park, PA, June 5–7, 2006). London: Taylor & Francis Group, pp. 361–366.

APPENDIX.—SUMMARY TABLES OF EXPLOSION PRESSURE DATA FOR THE LLEM TESTS

Table A-1.—Maximum explosion pressures during LLEM test #427

OMNIDIRECTIONAL PRESSURES AT THE WALL							
B-drift				C-drift			
Distance		Pressure		Distance		Pressure	
ft	m	psi	kPa	ft	m	psi	kPa
10	3	0.04	0.3	13	4	1.24	8.6
108	33	0.05	0.3	84	26	0.87	6.0
158	48	0.04	0.3	134	41	—	—
211	64	—	—	184	56	0.74	5.1
257	78	0.03	0.2	234	71	—	—
329	100	0.04	0.3	304	93	0.69	4.8
427	130	—	—	403	123	—	—
526	160	0.05	0.3	501	153	0.73	5.0
626	191	0.03	0.2	598	182	0.73	5.0
782	238	0.04	0.3	757	231	~0.71	~4.9

TOTAL EXPLOSION PRESSURES ON THE C-DRIFT SIDE STOPPINGS							
Location	Distance		Type	Pressure		Pressure-time integral, ∫Pdt	
	ft	m		psi	kPa	psi-s	kPa-s
Crosscut 4	355	108	Hollow-core block	0.74	5.1	0.51	3.5
Crosscut 5	451	138	Hollow-core block	—	—	—	—
Crosscut 6	547	167	Hollow-core block	—	—	—	—
Crosscut 7	647	197	Hollow-core block	0.73	5.0	0.54	3.7

Table A-2.—Maximum explosion pressures during LLEM test #428

OMNIDIRECTIONAL PRESSURES AT THE WALL							
B-drift				C-drift			
Distance		Pressure		Distance		Pressure	
ft	m	psi	kPa	ft	m	psi	kPa
10	3	0.6	4	13	4	5.1	35
108	33	0.4	3	84	26	4.5	31
158	48	0.3	2	134	41	—	—
211	64	—	—	184	56	4.4	30
257	78	0.3	2	234	71	—	—
329	100	0.4	2	304	93	3.8	26
427	130	—	—	403	123	3.4	24
526	160	—	—	501	153	3.4	24
626	191	0.4	2	598	182	2.9	20
782	238	0.5	4	757	231	2.7	19

TOTAL EXPLOSION PRESSURES ON THE C-DRIFT SIDE STOPPINGS							
Location	Distance		Type	Pressure		Pressure-time integral, ∫Pdt	
	ft	m		psi	kPa	psi-s	kPa-s
Crosscut 4	355	108	Hollow-core block[1]	5.2	36	0.51	3.5
Crosscut 5	451	138	Hollow-core block[1]	—	—	—	—
Crosscut 6	547	167	Hollow-core block	—	—	—	—
Crosscut 7	647	197	Hollow-core block	3.4	23	0.48	3.3

[1] Destroyed during test.

Table A-3.—Maximum explosion pressures during LLEM test #429

B-drift				C-drift			
Distance		Pressure		Distance		Pressure	
ft	m	psi	kPa	ft	m	psi	kPa
10	3	1.4	10	13	4	3.7	26
108	33	1.3	9	84	26	3.0	21
158	48	1.2	8	134	41	3.1	21
211	64	~1.1	~8	184	56	3.0	21
257	78	1.0	7	234	71	2.8	19
329	100	0.7	5	304	93	2.1	14
427	130	—	—	403	123	1.3	9
526	160	—	—	501	153	1.1	7
626	191	—	—	598	182	1.0	7
782	238	1.1	8	757	231	1.0	7

TOTAL EXPLOSION PRESSURES ON THE C-DRIFT SIDE STOPPINGS

Location	Distance		Type	Pressure		Pressure-time integral, ∫Pdt	
	ft	m		psi	kPa	psi-s	kPa-s
Crosscut 4	355	108	None	1.4	9	—	—
Crosscut 5	451	138	None	1.1	8	—	—
Crosscut 6	547	167	Hollow-core block	1.2	8	0.66	4.6
Crosscut 7	647	197	Hollow-core block	1.0	7	0.48	3.3

Table A-4.—Maximum explosion pressures during LLEM test #430

B-drift				C-drift			
Distance		Pressure		Distance		Pressure	
ft	m	psi	kPa	ft	m	psi	kPa
10	3	2.4	17	13	4	6.2	43
108	33	1.9	13	84	26	5.7	39
158	48	1.8	12	134	41	5.4	38
211	64	~1.9	~13	184	56	5.5	38
257	78	1.7	12	234	71	5.2	36
329	100	1.7	12	304	93	4.9	34
427	130	—	—	403	123	3.8	26
526	160	—	—	501	153	2.7	19
626	191	—	—	598	182	2.3	16
782	238	1.7	11	757	231	~1.7	~12

TOTAL EXPLOSION PRESSURES ON THE C-DRIFT SIDE STOPPINGS

Location	Distance		Type	Pressure		Pressure-time integral, ∫Pdt	
	ft	m		psi	kPa	psi-s	kPa-s
Crosscut 4	355	108	None	3.3	23	—	—
Crosscut 5	451	138	None	2.5	17	—	—
Crosscut 6	547	167	Hollow-core block[1]	3.8	26	0.86	5.9
Crosscut 7	647	197	Hollow-core block	2.2	15	0.66	4.6

[1]Three or four blocks knocked out near bottom of stopping during test.

Table A-5.—Maximum explosion pressures during LLEM test #432

OMNIDIRECTIONAL PRESSURES AT THE WALL

B-drift				C-drift			
Distance		Pressure		Distance		Pressure	
ft	m	psi	kPa	ft	m	psi	kPa
10	3	1.6	11	13	4	3.6	25
108	33	1.5	10	84	26	3.3	23
158	48	1.3	9	134	41	3.2	22
211	64	~1.1	~8	184	56	3.2	22
257	78	1.0	7	234	71	3.0	21
329	100	0.9	6	304	93	2.8	19
427	130	—	—	403	123	2.0	14
526	160	—	—	501	153	1.4	9
626	191	—	—	598	182	1.2	8
782	238	1.1	8	757	231	1.2	8

TOTAL EXPLOSION PRESSURES ON THE C-DRIFT SIDE STOPPINGS

Location	Distance		Type	Pressure		Pressure-time integral, ∫Pdt	
	ft	m		psi	kPa	psi-s	kPa-s
Crosscut 4	355	108	None	1.7	12	—	—
Crosscut 5	451	138	None	1.4	10	—	—
Crosscut 6	547	167	Hollow-core block[1]	1.7	12	0.61	4.2
Crosscut 7	647	197	Hollow-core block	1.2	9	0.51	3.5

[1]Two additional blocks knocked out near bottom of stopping during test.

Table A-6.—Maximum explosion pressures during LLEM test #433

OMNIDIRECTIONAL PRESSURES AT THE WALL

B-drift				C-drift			
Distance		Pressure		Distance		Pressure	
ft	m	psi	kPa	ft	m	psi	kPa
10	3	2.9	20	13	4	8.9	61
108	33	—	—	84	26	6.9	48
158	48	1.7	12	134	41	8.4	58
211	64	~1.8	~12	184	56	7.3	50
257	78	1.8	12	234	71	7.3	50
329	100	2.3	16	304	93	6.6	45
427	130	2.1	15	403	123	5.0	35
526	160	—	—	501	153	3.5	24
626	191	—	—	598	182	2.8	19
782	238	2.0	14	757	231	~1.9	~13

TOTAL EXPLOSION PRESSURES ON THE C-DRIFT SIDE STOPPINGS

Location	Distance		Type	Pressure		Pressure-time integral, ∫Pdt	
	ft	m		psi	kPa	psi-s	kPa-s
Crosscut 4	355	108	None	4.2	29	—	—
Crosscut 5	451	138	None	3.1	21	—	—
Crosscut 6	547	167	Hollow-core block[1]	3.6	25	0.91	6.3
Crosscut 7	647	197	Hollow-core block	2.8	19	0.64	4.4

[1]Destroyed during test.

Table A-7.—Maximum explosion pressures during LLEM test #434

OMNIDIRECTIONAL PRESSURES AT THE WALL

B-drift				C-drift			
Distance		Pressure		Distance		Pressure	
ft	m	psi	kPa	ft	m	psi	kPa
10	3	2.8	19	13	4	7.6	53
108	33	—	—	84	26	6.7	46
158	48	2.7	18	134	41	7.4	51
211	64	3.0	21	184	56	6.4	44
257	78	2.8	19	234	71	6.7	47
329	100	2.2	15	304	93	5.1	35
427	130	2.2	15	403	123	4.1	29
526	160	—	—	501	153	2.9	20
626	191	—	—	598	182	2.4	16
782	238	2.0	14	757	231	2.3	16

TOTAL EXPLOSION PRESSURES ON THE C-DRIFT SIDE STOPPINGS

Location	Distance		Type	Pressure		Pressure-time integral, ∫Pdt	
	ft	m		psi	kPa	psi-s	kPa-s
Crosscut 4	355	108	None	3.4	23	—	—
Crosscut 5	451	138	None	2.6	18	—	—
Crosscut 6	547	167	None	2.9	20	—	—
Crosscut 7	647	197	Hollow-core block[1]	2.3	16	3.0	21

[1]Destroyed during test.

Table A-8.—Maximum explosion pressures during LLEM test #457

OMNIDIRECTIONAL PRESSURES AT THE WALL

B-drift				C-drift			
Distance		Pressure		Distance		Pressure	
ft	m	psi	kPa	ft	m	psi	kPa
10	3	<0.1	<1	13	4	1.6	11
108	33	<0.1	<1	84	26	0.8	6
158	48	<0.1	<1	134	41	0.6	4
211	64	<0.2	<1	184	56	0.8	5
257	78	<0.1	<1	234	71	0.7	5
329	100	<0.1	<1	304	93	0.7	5
427	130	<0.1	<1	403	123	0.7	5
526	160	<0.1	<1	501	153	0.7	5
626	191	<0.1	<1	598	182	0.8	5
782	238	<0.1	<1	757	231	0.7	5

TOTAL EXPLOSION PRESSURES ON THE C-DRIFT SIDE STOPPINGS

Location	Distance		Type	Pressure		Pressure-time integral, ∫Pdt	
	ft	m		psi	kPa	psi-s	kPa-s
Crosscut 4	355	108	Solid block	0.78	5.4	0.51	3.5
Crosscut 5	451	138	Solid block	0.75	5.2	0.53	3.7
Crosscut 6	547	167	Steel panel	0.81	5.6	0.55	3.8
Crosscut 7	647	197	Steel panel	0.75	5.2	0.55	3.8

Table A-9.—Maximum explosion pressures during LLEM test #458

OMNIDIRECTIONAL PRESSURES AT THE WALL							
B-drift				C-drift			
Distance		Pressure		Distance		Pressure	
ft	m	psi	kPa	ft	m	psi	kPa
10	3	0.6	4	13	4	1.9	13
108	33	0.6	4	84	26	1.6	11
158	48	0.5	3	134	41	1.1	8
211	64	~0.4	~3	184	56	1.5	10
257	78	~0.3	~2	234	71	1.4	9
329	100	0.4	3	304	93	1.4	9
427	130	0.4	3	403	123	1.4	10
526	160	0.4	3	501	153	1.2	8
626	191	0.2	1	598	182	—	—
782	238	0.2	1	757	231	1.2	8

TOTAL EXPLOSION PRESSURES ON THE C-DRIFT SIDE STOPPINGS							
Location	Distance		Type	Pressure		Pressure-time integral, ∫Pdt	
	ft	m		psi	kPa	psi-s	kPa-s
Crosscut 4	355	108	Solid block	1.6	11	0.51	3.5
Crosscut 5	451	138	Solid block	1.5	10	0.38	2.6
Crosscut 6	547	167	Steel panel[1]	1.3	9	0.30	2.0
Crosscut 7	647	197	Steel panel[1]	1.3	9	0.30	2.0

[1]Failed during test.

Table A-10.—Maximum explosion pressures during LLEM test #459

OMNIDIRECTIONAL PRESSURES AT THE WALL							
B-drift				C-drift			
Distance		Pressure		Distance		Pressure	
ft	m	psi	kPa	ft	m	psi	kPa
10	3	0.2	1	13	4	4.3	30
108	33	0.2	1	84	26	3.5	24
158	48	0.2	1	134	41	2.5	17
211	64	0.2	1	184	56	3.8	26
257	78	0.2	1	234	71	3.5	24
329	100	0.1	1	304	93	3.4	23
427	130	0.2	1	403	123	2.8	19
526	160	0.3	2	501	153	2.7	19
626	191	0.4	2	598	182	2.3	16
782	238	0.4	2	757	231	2.4	17

TOTAL EXPLOSION PRESSURES ON THE C-DRIFT SIDE STOPPINGS							
Location	Distance		Type	Pressure		Pressure-time integral, ∫Pdt	
	ft	m		psi	kPa	psi-s	kPa-s
Crosscut 4	355	108	Solid block	4.6	32	0.47	3.3
Crosscut 5	451	138	Solid block	3.4	23	0.48	3.3
Crosscut 6	547	167	Australian woven cloth	3.0	21	0.54	3.7
Crosscut 7	647	197	Australian woven cloth	2.8	19	0.55	3.8

Table A-11.—Maximum explosion pressures during LLEM test #460

OMNIDIRECTIONAL PRESSURES AT THE WALL							
B-drift				C-drift			
Distance		Pressure		Distance		Pressure	
ft	m	psi	kPa	ft	m	psi	kPa
10	3	0.3	2	13	4	6.3	43
108	33	0.2	1	84	26	5.8	40
158	48	0.2	1	134	41	5.6	39
211	64	0.2	1	184	56	5.5	38
257	78	0.2	1	234	71	5.2	36
329	100	0.2	1	304	93	5.0	34
427	130	0.2	2	403	123	4.0	27
526	160	0.3	2	501	153	3.7	26
626	191	0.4	3	598	182	3.4	23
782	238	0.3	2	757	231	3.1	22

TOTAL EXPLOSION PRESSURES ON THE C-DRIFT SIDE STOPPINGS							
Location	Distance		Type	Pressure		Pressure-time integral, ∫Pdt	
	ft	m		psi	kPa	psi-s	kPa-s
Crosscut 4	355	108	Solid block	6.7	46	0.57	3.9
Crosscut 5	451	138	Solid block	4.7	33	0.56	3.8
Crosscut 6	547	167	Australian woven cloth	4.0	27	0.50	3.4
Crosscut 7	647	197	Australian woven cloth	3.9	27	0.50	3.5

Table A-12.—Maximum explosion pressures during LLEM test #461

OMNIDIRECTIONAL PRESSURES AT THE WALL							
B-drift				C-drift			
Distance		Pressure		Distance		Pressure	
ft	m	psi	kPa	ft	m	psi	kPa
10	3	0.2	1	13	4	NA	NA
108	33	0.1	1	84	26	4.1	28
158	48	0.1	1	134	41	4.3	30
211	64	0.2	1	184	56	4.0	28
257	78	0.2	1	234	71	4.0	28
329	100	0.1	1	304	93	3.4	23
427	130	0.1	1	403	123	3.5	24
526	160	0.2	1	501	153	3.2	22
626	191	0.2	1	598	182	3.2	22
782	238	0.1	1	757	231	3.2	22

TOTAL EXPLOSION PRESSURES ON THE C-DRIFT SIDE STOPPINGS							
Location	Distance		Type	Pressure		Pressure-time integral, ∫Pdt	
	ft	m		psi	kPa	psi-s	kPa-s
Crosscut 4	355	108	Solid block	4.2	29	1.32	9.1
Crosscut 5	451	138	Solid block	4.0	28	1.18	8.1
Crosscut 6	547	167	Australian woven cloth	3.7	26	1.18	8.1
Crosscut 7	647	197	Australian woven cloth	3.6	25	1.19	8.2

NA Not available.

Table A-13.—Maximum explosion pressures during LLEM test #462

OMNIDIRECTIONAL PRESSURES AT THE WALL

B-drift				C-drift			
Distance		Pressure		Distance		Pressure	
ft	m	psi	kPa	ft	m	psi	kPa
10	3	0.9	6	13	4	NA	NA
108	33	0.8	6	84	26	6.5	44
158	48	0.7	5	134	41	6.8	47
211	64	0.6	4	184	56	6.1	42
257	78	0.5	4	234	71	6.4	44
329	100	0.8	5	304	93	5.5	38
427	130	0.9	6	403	123	5.8	40
526	160	1.1	7	501	153	5.2	36
626	191	0.8	6	598	182	4.9	34
782	238	0.7	5	757	231	4.8	33

TOTAL EXPLOSION PRESSURES ON THE C-DRIFT SIDE STOPPINGS

Location	Distance		Type	Pressure		Pressure-time integral, $\int Pdt$	
	ft	m		psi	kPa	psi-s	kPa-s
Crosscut 4	355	108	Solid block[1]	7.6	52	1.04	7.2
Crosscut 5	451	138	Solid block	6.7	46	1.06	7.3
Crosscut 6	547	167	Australian woven cloth[1]	6.1	42	0.71	4.9
Crosscut 7	647	197	Australian woven cloth[1]	5.4	37	0.85	5.9

NA Not available. [1]Destroyed during test.

Table A-14.—Maximum explosion pressures during LLEM test #463

OMNIDIRECTIONAL PRESSURES AT THE WALL

B-drift				C-drift			
Distance		Pressure		Distance		Pressure	
ft	m	psi	kPa	ft	m	psi	kPa
10	3	3.9	27	13	4	NA	NA
108	33	3.5	24	84	26	17.7	122
158	48	3.3	23	134	41	18.1	125
211	64	3.0	20	184	56	20.0	138
257	78	2.8	19	234	71	15.1	104
329	100	4.6	32	304	93	16.6	114
427	130	4.6	32	403	123	12.4	85
526	160	4.3	30	501	153	9.6	66
626	191	3.6	25	598	182	7.2	50
782	238	2.8	20	757	231	5.4	37

TOTAL EXPLOSION PRESSURES ON THE C-DRIFT SIDE STOPPINGS

Location	Distance		Type	Pressure		Pressure-time integral, $\int Pdt$	
	ft	m		psi	kPa	psi-s	kPa-s
Crosscut 4	355	108	None	11.6	80	NA	NA
Crosscut 5	451	138	Solid block[1]	16.6	115	1.34	9.2
Crosscut 6	547	167	None	7.2	50	NA	NA
Crosscut 7	647	197	None	5.9	41	NA	NA

NA Not available. [1]Destroyed during test.

Table A-15.—Maximum explosion pressures during LLEM test #491

OMNIDIRECTIONAL PRESSURES AT THE WALL							
B-drift				C-drift			
Distance		Pressure		Distance		Pressure	
ft	m	psi	kPa	ft	m	psi	kPa
10	3	5.5	38	13	4	6.2	43
108	33	3.2	22	84	26	4.4	30
158	48	3.8	26	134	41	3.7	26
211	64	3.4	23	184	56	3.5	24
257	78	4.2	29	234	71	3.6	25
329	100	3.7	25	304	93	2.8	19
427	130	3.1	21	403	123	3.3	23
526	160	2.3	16	501	153	2.4	17
626	191	—	—	598	182	2.0	14
782	238	2.2	15	757	231	1.9	13

TOTAL EXPLOSION PRESSURES ON THE B-DRIFT SIDE STOPPINGS							
Location	Distance		Type[1]	Pressure		Pressure-time integral, ∫Pdt	
	ft	m		psi	kPa	psi-s	kPa-s
Crosscut 3AB	281	86	Hollow-core block	6.4	44	1.4	9.4
Crosscut 3BC	281	86	Hollow-core block	—	—	—	—
Crosscut 4AB	377	115	Hollow-core block	4.0	28	1.2	8.1
Crosscut 4BC	377	115	Hollow-core block	3.6	25	—	—

NOTE.—Ignition at ~221 ft (~67 m) from closed end of B-drift.
[1] All of the stoppings were destroyed during the test.

Table A-16.—Maximum explosion pressures during LLEM test #494

OMNIDIRECTIONAL PRESSURES AT THE WALL							
B-drift				C-drift			
Distance		Pressure		Distance		Pressure	
ft	m	psi	kPa	ft	m	psi	kPa
10	3	0.40	2.8	13	4	0.40	2.8
108	33	0.40	2.8	84	26	0.38	2.6
158	48	0.39	2.7	134	41	0.36	2.5
211	64	0.40	2.8	184	56	0.35	2.4
257	78	0.35	2.4	234	71	0.35	2.4
329	100	0.39	2.7	304	93	0.28	1.9
427	130	0.48	3.3	403	123	0.29	2.0
526	160	0.27	1.9	501	153	0.29	2.0
626	191	0.28	1.9	598	182	0.29	2.0
782	238	—	—	757	231	0.25	1.7

TOTAL EXPLOSION PRESSURES ON THE B-DRIFT SIDE STOPPINGS							
Location	Distance		Type	Pressure		Pressure-time integral, ∫Pdt	
	ft	m		psi	kPa	psi-s	kPa-s
Crosscut 3AB	281	86	Hollow-core block	0.37	2.6	0.37	2.6
Crosscut 3BC	281	86	Hollow-core block	0.40	2.8	0.39	2.7
Crosscut 4AB	377	115	Hollow-core block	0.44	3.0	0.40	2.8
Crosscut 4BC	377	115	Hollow-core block	0.44	3.0	0.40	2.8
B-drift	446	136	Hollow-core block with regulator	0.49	3.4	0.45	3.1

NOTE.—Ignition at ~250 ft (~76 m) from closed end of B-drift.

Table A-17.—Maximum explosion pressures during LLEM test #495

OMNIDIRECTIONAL PRESSURES AT THE WALL							
B-drift				C-drift			
Distance		Pressure		Distance		Pressure	
ft	m	psi	kPa	ft	m	psi	kPa
10	3	0.75	5.2	13	4	0.85	5.9
108	33	0.67	4.6	84	26	0.74	5.1
158	48	0.63	4.3	134	41	0.62	4.3
211	64	0.78	5.4	184	56	0.56	3.9
257	78	0.82	5.7	234	71	0.56	3.9
329	100	0.80	5.5	304	93	0.53	3.7
427	130	1.21	8.3	403	123	0.44	3.0
526	160	0.45	3.1	501	153	0.47	3.2
626	191	0.42	2.9	598	182	0.46	3.2
782	238	—	—	757	231	0.43	3.0

TOTAL EXPLOSION PRESSURES ON THE B-DRIFT SIDE STOPPINGS							
Location	Distance		Type	Pressure		Pressure-time integral, ∫Pdt	
	ft	m		psi	kPa	psi-s	kPa-s
Crosscut 3AB	281	86	Hollow-core block	0.89	6.1	0.49	.04
Crosscut 3BC	281	86	Hollow-core block	0.96	6.6	0.53	3.7
Crosscut 4AB	377	115	Hollow-core block	1.08	7.5	0.53	3.7
Crosscut 4BC	377	115	Hollow-core block	1.06	7.3	0.53	3.7
B-drift	446	136	Hollow-core block with regulator	1.24	8.6	0.54	3.7

NOTE.—Ignition at ~250 ft (~76 m) from closed end of B-drift.

Table A-18.—Maximum explosion pressures during LLEM test #496

OMNIDIRECTIONAL PRESSURES AT THE WALL							
B-drift				C-drift			
Distance		Pressure		Distance		Pressure	
ft	m	psi	kPa	ft	m	psi	kPa
10	3	2.5	17	13	4	3.1	21
108	33	1.6	11	84	26	2.1	15
158	48	1.9	13	134	41	1.9	13
211	64	2.3	16	184	56	1.7	12
257	78	2.4	16	234	71	1.7	12
329	100	2.2	15	304	93	1.6	11
427	130	3.5	24	403	123	1.5	10
526	160	1.0	7	501	153	1.2	8
626	191	0.9	6	598	182	1.1	8
782	238	—	—	757	231	1.0	7

TOTAL EXPLOSION PRESSURES ON THE B-DRIFT SIDE STOPPINGS							
Location	Distance		Type	Pressure		Pressure-time integral, ∫Pdt	
	ft	m		psi	kPa	psi-s	kPa-s
Crosscut 3AB	281	86	Hollow-core block	2.8	19	0.53	3.7
Crosscut 3BC	281	86	Hollow-core block	2.9	20	0.53	3.7
Crosscut 4AB	377	115	Hollow-core block	3.1	21	0.84	5.8
Crosscut 4BC	377	115	Hollow-core block	3.0	21	0.83	5.7
B-drift	446	136	Hollow-core block with regulator	3.5	24	0.87	6.0

NOTE.—Ignition at ~250 ft (~76 m) from closed end of B-drift.

Table A-19.—Maximum explosion pressures during LLEM test #497

OMNIDIRECTIONAL PRESSURES AT THE WALL							
B-drift				C-drift			
Distance		Pressure		Distance		Pressure	
ft	m	psi	kPa	ft	m	psi	kPa
10	3	4.2	29	13	4	4.0	28
108	33	2.7	19	84	26	3.2	22
158	48	2.4	17	134	41	2.7	19
211	64	2.5	17	184	56	2.6	18
257	78	3.4	23	234	71	2.6	18
329	100	3.1	21	304	93	2.4	17
427	130	5.1	35	403	123	2.1	15
526	160	1.7	12	501	153	1.8	12
626	191	1.6	11	598	182	1.6	11
782	238	1.6	11	757	231	1.4	10

TOTAL EXPLOSION PRESSURES ON THE B-DRIFT SIDE STOPPINGS						
Location	Distance		Type	Pressure		Pressure-time integral, ∫Pdt	
	ft	m		psi	kPa	psi-s	kPa-s
Crosscut 3AB	281	86	Hollow-core block	4.3	30	0.70	4.8
Crosscut 3BC	281	86	Hollow-core block[1]	4.2	29	0.71	4.9
Crosscut 4AB	377	115	Hollow-core block[2]	3.8	26	~0.94	~6.5
Crosscut 4BC	377	115	Hollow-core block[2]	3.6	25	0.93	6.4
B-drift	446	136	Hollow-core block with regulator[2]	5.2	36	0.97	6.7

NOTE.—Ignition at ~221 ft (~67 m) from closed end of B-drift.
[1]Partially damaged during test.
[2]Destroyed during test.

Table A-20.—Maximum explosion pressures during LLEM test #510

OMNIDIRECTIONAL PRESSURES AT THE WALL							
A-drift				B-drift			
Distance		Pressure		Distance		Pressure	
ft	m	psi	kPa	ft	m	psi	kPa
0	0	14.6	101	10	3	0.7	5
22	7	14.2	98	108	33	—	
81	25	13.5	93	158	48	—	
132	40	14.0	96	211	64	0.3	2
183	56	12.4	86	257	78	0.3	2
233	71	13.6	93	329	100	0.3	2
283	86	12.6	87	427	130	0.3	2
355	108	12.2	84	526	160	0.4	3
453	138	11.6	80	626	191	0.6	4
550	168	10.8	74	782	238	1.0	7
649	198	10.7	74				
807	246	10.3	71				

TOTAL EXPLOSION PRESSURES ON THE A-DRIFT SIDE STOPPINGS						
Location	Distance		Type	Pressure		Pressure-time integral, ∫Pdt	
	ft	m		psi	kPa	psi-s	kPa-s
Crosscut 6	600	183	8-in (20-cm) thick wet-laid solid concrete block	14.5	100	3.0	21
Crosscut 7	699	213	6-in (15-cm) thick wet-laid solid concrete block	13.5	93	3.0	21

Table A-21.—Maximum explosion pressures during LLEM test #512

OMNIDIRECTIONAL PRESSURES AT THE WALL							
A-drift				B-drift			
Distance		Pressure		Distance		Pressure	
ft	m	psi	kPa	ft	m	psi	kPa
0	0	13.2	91	10	3	0.6	4
22	7	13.1	90	108	33	—	—
81	25	12.5	86	158	48	—	—
132	40	11.8	82	211	64	0.3	2
183	56	10.3	71	257	78	0.3	2
233	71	10.7	74	329	100	0.3	2
283	86	10.0	69	427	130	0.3	2
355	108	8.3	57	526	160	0.4	3
453	138	8.4	58	626	191	0.5	3
550	168	8.1	56	782	238	0.8	6
649	198	8.2	57				
807	246	8.4	58				

TOTAL EXPLOSION PRESSURES ON THE A-DRIFT SIDE STOPPINGS							
Location	Distance		Type	Pressure		Pressure-time integral, ∫Pdt	
	ft	m		psi	kPa	psi-s	kPa-s
Crosscut 6	600	183	8-in (20-cm) thick wet-laid solid concrete block	10.4	72	4.6	32
Crosscut 7	699	213	6-in (15-cm) thick wet-laid solid concrete block	10.1	69	4.4	31

Table A-22.—Maximum explosion pressures during LLEM test #515

OMNIDIRECTIONAL PRESSURES AT THE WALL							
A-drift				B-drift			
Distance		Pressure		Distance		Pressure	
ft	m	psi	kPa	ft	m	psi	kPa
0	0	13.4	93	10	3	0.8	6
22	7	13.3	91	108	33	—	—
81	25	13.0	90	158	48	—	—
132	40	14.0	96	211	64	0.3	2
183	56	12.7	88	257	78	0.4	2
233	71	13.9	96	329	100	0.3	2
283	86	12.7	88	427	130	0.4	3
355	108	11.9	82	526	160	0.5	4
453	138	~12	~83	626	191	0.6	4
550	168	11.5	79	782	238	1.1	8
649	198	11.7	80				
807	246	11.1	76				

TOTAL EXPLOSION PRESSURES ON THE A-DRIFT SIDE STOPPINGS							
Location	Distance		Type	Pressure		Pressure-time integral, ∫Pdt	
	ft	m		psi	kPa	psi-s	kPa-s
Crosscut 6	600	183	8-in (20-cm) thick wet-laid solid concrete block	14.2	98	3.9	27
Crosscut 7	699	213	6-in (15-cm) thick wet-laid solid concrete block	14.2	98	3.9	27

Table A-23.—Maximum explosion pressures during LLEM test #519

OMNIDIRECTIONAL PRESSURES AT THE WALL							
A-drift				B-drift			
Distance		Pressure		Distance		Pressure	
ft	m	psi	kPa	ft	m	psi	kPa
0	0	24.8	171	10	3	4.7	32
22	7	23.8	164	108	33	—	—
81	25	22.1	152	158	48	—	—
132	40	23.7	163	211	64	2.7	18
183	56	21.7	150	257	78	2.5	17
233	71	24.9	172	329	100	2.5	17
283	86	21.5	148	427	130	2.0	14
355	108	20.9	144	526	160	2.1	14
453	138	~20	~135	626	191	2.6	18
550	168	19.3	133	782	238	2.6	18
649	198	19.6	135				
807	246	17.6	122				

TOTAL EXPLOSION PRESSURES ON THE A-DRIFT SIDE STOPPINGS							
Location	Distance		Type	Pressure		Pressure-time integral, ∫Pdt	
	ft	m		psi	kPa	psi-s	kPa-s
Crosscut 6	600	183	8-in (20-cm) thick wet-laid solid concrete block	26.8	185	4.2	29
Crosscut 7	699	213	6-in (15-cm) thick wet-laid solid concrete block[1]	25.3	174	3.2	22

[1]Destroyed during test.